America's Struggle for
Leadership in Technology

America's Struggle for Leadership in Technology

Jean-Claude Derian

translated by Severen Schaeffer

The MIT Press
Cambridge, Massachusetts
London, England

©1990 Éditions Albin Michel

Printed and bound in the United States of America.

Published in Paris by Éditions Albin Michel as *La Grande Panne de la Technologie Américaine* (1988).

Library of Congress Cataloging-in-Publication Data

Derian, Jean-Claude.
  [Grande panne de la technologie américaine. English]
  America's struggle for leadership in technology / Jean-Claude
Derian.
    p.   cm.
  Translation of: La grande panne de la technologie américaine.
  ISBN 0-262-04102-2
  1. Technology—United States.   I. Title.
  T21.D4713   1990
  338.97306—dc20   89-27972
                                                           CIP

Contents

Foreword

Success in technology seemed to come easy in America. The American consumer has always rewarded novelty and admired progress. Our admiration of individual creativity and entrepreneurship found ample gratification. We Americans were—and are—very good at scientific research and great engineering adventures. After 1946, we were good because we could afford it. We had the good sense to put the scientists in charge of "big science" projects. They selected their projects with great care and executed them with almost uniform success, drawing on the skills of colleagues from all over the world. Our politicians committed their futures to a few big engineering spectaculars, such as the Apollo program, and provided the funds needed to succeed.

With the advent of the 1980s came a series of serious shocks to this macho technical self-image. American high-tech firms began losing market share to German and Japanese firms. The *Challenger* disaster was followed by a series of launch-vehicle failures. Confusion surrounded the claims of advocates and critics of strategic defense as to the merits of a technological fix for the nuclear nightmare. The fact that none of these projects for big science facilities or for adventures promoting national self-esteem had much to do with industrial innovation or productivity was beside the point. Government big science became the surrogate for national technical prowess.

Then came the budget deficit. Although the deficit is only partly a reflection of shortcomings in US technological performance, political gridlock holds defense, entitlements, and interest on the debt to be sacrosanct. Thus, for the first time, federal research and development expenditures constitute a substantial fraction (some estimate 27 percent) of the discretionary part of the federal budget. The time has come for hard choices.

Yet at no other time in recent memory have so many new technological "megaprojects" been initiated by an administration. The *Challenger* replacement, the space station, the superconducting supercollider, the program to sequence and map the human genome, and the aerospace plane—each one a multi-billion-dollar commitment—were all launched by the Reagan administration into this sea of red ink.

In addition, American technological self-confidence has been damaged, in part because industrial competitors often invest more technical talent in areas to which American engineering schools accord low prestige: design and manufacturing. This does seem a strange time to look to heroic government-sponsored projects, especially when there may not be enough money or will to ensure success.

In times like these we need an objective look from the outside, some helpful advice from a well-informed good friend. Along comes Jean-Claude Derian to tell us that our technology system is not irreparably broken but is seriously out of whack.

I like the simile of technology as a mirage. To Derian the mirage is something promising that becomes increasingly inaccessible—the brass ring that always remains just out of reach. But Americans have grasped that ring and flung it around the moon. We built the industries that set the target for others to challenge.

The American problem is not that the technology remains out of reach. The problem is that the government's technology strategies and the management of the program have lacked precision, integrity, and discipline. They have been built out of images of an impenetrable space shield, of two-hour flights between New York and Tokyo, of permanent habitats or factories in space—in short, mirages.

Perhaps a more appropriate simile is the mirage as oasis—the apparent but unreal solution to all our problems. The mirage of a distant watering hole only distracts us from the need to search for water and shade where we are. Our mirage is not out of reach; it is simply not relevant to the primary source of our predicament, which has much more to do with public profligacy and private mismanagement than with technological capability.

A French perspective on the American technological scene could be particularly appropriate under these circumstances. If the French approach to managing big government technology programs has shortcomings, competent and disciplined administration is not one of them. The Grandes Ecoles produce an elite corps of engineers who enter public service and who produce power reactors, high-speed trains, deep-ocean explorers, space rockets, and satellites of admirable capability.

Jean-Claude Derian has an important message for Americans: the "new American challenge" to Europe proclaimed by Jean-Jacques Servan-Schreiber seems to have fallen short under the competitive onslaught from Asia and Europe (and also because of an overpriced dollar).

Why? Derian presents a model that is centered on the notion of two different technical cultures. This is a model with particular appeal to Europe, as it draws on a long history of admiration for competitive American entrepreneurship and of familiarity with conservative, established European enterprise enjoying the patronage of the state.

The first of Derian's two cultures is the "exposed" culture of competitive commercial activity, of government support for basic science and higher education in support of innovative new ventures: Silicon Valley, Route 128, and the Research Triangle. The second is the "sheltered" culture of high-technology multinational companies whose growth was ensured by the protection of a vast domestic market and by support through military and space procurement (or, in the case of telecommunications, by the warm blanket of status as a public monopoly).

Derian's thesis is that, whereas the Japanese are very competitive in the exposed culture, the capabilities of Europe (especially France and the United Kingdom) are concentrated in the sheltered culture. It is doubtless true that space and defense megaprojects draw American companies away from many of the skills most critical to international commercial competition in mass-produced products. The mirage of "spinoff" technology has lured many an enterprise over the dunes into a desert of troubles.

He is optimistic—at least from the US point of view—that deregulation will make the sheltered sector more competitive and that a revival of US manufacturing technology will make the exposed sector more competitive too. Perhaps the US government will also learn to shift its own technology strategy toward support for industry initiatives (Sematech seems the prototype) and away from projects with questionable goals (such as the National Aerospace Plane).

This book provides food for thought not only for Americans searching for new public-private relationships role in support of competitiveness but for Europeans as well.

The challenge to all industrial nations is to find the means to make both sheltered and exposed industrial cultures efficient, creative, and responsive to public needs. Europe, more proficient that the United States at the provision of infrastructure and more comfortable with the sheltering of technology through government promotion, will press forward with a more competitive environment after the 1992 economic integration.

Japan's best firms are already very competitive in cost and quality, and in speed to market with new technology. But they are also organized into very large, conservatively managed corporate empires, fiercely competitive domestically but sheltered by the state in their pursuit of world markets.

In the United States, the Reagan and Bush administrations have responded to the double challenge of strong competition in international trade and crumbling infrastructure and public services with more exposure through deregulation and open markets, and with faith in private entrepreneurship. But many Americans doubt that this strategy will be enough. The United States is experimenting with new roles for government: a new Technology Administration in the Department of Commerce, and bootleg industrial-policy projects in the Department of Defense.

Derian's warning should be heeded. Perhaps we should spend fewer resources searching for the oasis and put more concerted effort into nurturing the aquifer of the national science and technology base that motivates our universities, supports public services, and nourishes our industry.

Lewis M. Branscomb
Harvard University

Acknowledgments

This book is the result of my observations, discussions, and on-site visits over a period of three years spent in a privileged position for observing and being in touch with the American scientific, technological, and industrial world, as head of the Scientific Mission at the French Embassy in the United States. I could never have accomplished this work without the invaluable assistance of all the members of the Mission, with whom I exchanged ideas and information and who gave support in their respective specialties. I offer my warmest thanks to all. Thanks, in particular, to Karen Daifuku, the Mission's documentalist, for her efficient help in collecting the documents and the bibliographic references.

This work also owes a great deal to the American university professors, government officials, industrial managers, and consultants with whom I held stimulating conversations on various aspects of the subject. My thanks in particular go to Harvard professors Harvey Brooks and Lewis Branscomb, Stanford professor Nathan Rosenberg, Berkeley professor John Zysman, MIT professor Eugene Skolnikoff, General Electric vice-president for research Roland Schmitt, and MITRE Corporation president Charles Zraket.

I would like to express my appreciation to all those, both French and American, who were kind enough to read parts of the manuscript and to share their remarks concerning their specialties and interests: Henri Blanc, Irvin Bupp, Jean-Marc de Comarmond, Alain Crémieux, Hubert Curien, Herbert Fusfeld, Jacqueline Grapin, Dan Greenberg, Pierre-Henri Gourgeon, Jean-Claude Hirel, Peter Huber, Louis Laidet, François de Lavergne, Francis Latapie, Albert Lumbroso, Claude Mandil, Laurence Ratier-Coutrot, Walter Rosenblith, Christian Sautter, Jacques Tamisier, and Alain Touraine.

Jean-Claude Derian

America's Struggle for
Leadership in Technology

Introduction to American Edition

More than a century ago, in *Democracy in America*, Tocqueville was the first of my compatriots to look at America as an outsider and discuss the merits of the American political system—an approach that turned out to be most fruitful and that has since then been followed many times.

Though I am no Tocqueville, I too have taken a "foreign eye" approach in writing about America. I was posted to Washington for three years as the science counselor at the Embassy of France, and I think I learned a few things about science and technology in the United States. This book is the product of this learning experience, and therefore it represents a French view—although a personal one—of the American high-technology system. Since it was first written for a French audience, some chapters contain details on the US system that American readers might consider unnecessary. I hope that they will forgive me and simply skip over these parts.

The period when I was in Washington was an especially interesting one for the subject of this book. The summer of 1984, when I arrived, was close to the end of Ronald Reagan's first term. It was a time of certainty and triumph for the Great Communicator. The economy was back on track, unemployment was declining sharply, and an unprecedented effort was underway to reshape America's military power—a necessary condition, the president had said when he was elected four years earlier, for restoring the nation's strength and influence with respect to the Soviet Union.

The Strategic Defense Initiative, launched a year earlier, was to crown this ambition. Enthusiasm ran high in the military and the space industry for exploring this new frontier—a challenge that had created perplexity if not concern among Europeans, who feared an even wider technological gap between Europe and the United States. Moreover, SDI was only one of several ambitious technology-oriented projects launched by Ronald Reagan. By the end of 1984 he had endorsed the National Aeronautics and Space Administration's proposal to build a permanently inhabited space station. Then came the National Aerospace Plane, nicknamed "the Orient Express" since it would shorten to two hours the trip from New York to Tokyo. Later came the Superconducting Supercollider, the largest machine ever envisioned for exploring the frontiers of knowledge in elementary particles. If we include several major military projects, such as the stealth bomber, never before in history had such a large number of technology-oriented programs been launched in such a short period of time as under Ronald Reagan's presidency.

It is ironic, though, that as these extraordinarily ambitious goals were being assigned to technology by the federal government, American high-tech[1] companies were facing unprecedented hard times. The 1980s had witnessed the overwhelming superiority of Japanese companies in many high-tech markets, such as semiconductors, new materials, and telecommunications equipment. They had also seen a number of European companies penetrate US internal markets. The most striking examples are Airbus Industrie in civilian aeronautics, Siemens in fiber optics and medical imaging, and the Ariane rocket, which now dominates the market for launching commercial satellites.

Yet America in the mid-1980s was still innovating with the same feverish energy as in the 1960s. The discovery in 1986 of high-temperature superconductivity mobilized universities and industry in a frantic race toward this new frontier of knowledge. But the United States seemed increasingly less efficient in exploiting its tremendous creative capacity. The United States was still the world's leader in research and innovation, but Japan, Taiwan, and (in some domains) the European countries were flooding the American high-tech markets with their products. In 1986, for the first time in its history, the United States showed a trade deficit in high-technology sectors.

It looked as though technology had become a mirage for America. Never before in the history of the United States had the gap been so wide between technology-based hopes and expectations and real scientific and technological progress. Will America be able to escape being misled by this mirage? The first half of this book examines this question, and the last two chapters suggest an answer. The analysis turns on the hypothesis that American high-tech industrial development results from a blend of two different kinds of technical cultures, two different innovation processes, two different types of companies, and two different kinds of foreign competitors.

The "exposed culture," which is the heritage of two centuries of invention and creativity in American industry, is individualistic and entrepreneurial. This technical culture stems from the permanent confrontation of ideas, innovations, and new products and processes in the marketplace. Most of the time it characterizes the production and the marketing, in a highly competitive environment, of standardized goods produced for a large number of customers. This highly competitive and selective innovation process has produced Route 128 and Silicon Valley. But today it is challenged by an apparently more efficient approach to innovation coming from the Far East.

The "sheltered culture" has resulted from what seems to be an accident of history: the involvement of the federal government in the generation of advanced military technology in the aftermath of World War II. Although it has varied over time, this involvement has turned out to be permanent and has represented the most formidable financing ever allocated to technological progress. This progress has, however, been driven by very specific forces and needs. High sophistication, growing technical complexity, and products custom-made for a single customer or a limited number of customers are the most salient characteristics of this technical culture. Often influenced if not dominated by the governmental's needs, this technical culture has also led over time to the development of large commercial non-government markets, such as those for civilian jetliners, large computers, and telecommunication switching equipment. Whether they are supplying the government or private markets, these industries have an important characteristic in common: until the early 1980s, companies within them enjoyed the protection of a monopoly or an oligopoly, of large-scale government procurement, or of other types of protection that limited the entry of outside competitors. Hence the name "sheltered culture." Like companies belonging to the exposed culture, US firms in the sheltered culture are today being challenged by foreign competition. However, this challenge does not come from Japan or Taiwan but from Europe. For example, Airbus Industrie, the European consortium for aeronautics, has brought international competition into the universe of the American sheltered culture for the first time in history.

But there are hopeful signs. Chapter 11 reviews America's new comparative advantages for a comeback in the technology race. A lower exchange rate for the dollar since 1985, growing trade tensions and a worldwide tendency toward more protectionism, and the availability of a new generation of computer-integrated manufacturing technologies are creating new incentives for multinational corporations—American, Japanese, and European—to manufacture in the United States. Some of these factors are also helping smaller American companies produce goods in the United States at competitive prices.

In the sheltered sector, deregulation initiated by the US government in the late 1970s brutally opened to competition many markets that were previously protected. As will be shown in chapter 10, this change has been particularly beneficial to foreign companies. After a period of shock, however, most US firms are progressively adjusting to the new rules; this is true for companies as different as Boeing and AT&T, in spite of their substantial losses of market shares.

Lessons from the Recent Past

My optimistic conclusions, written in early 1988, may be viewed as somewhat premature. Recent developments suggest that the technological revival in America remains to be definitively confirmed, for recent events have brought both good and bad news for American high tech.

Among the disturbing pieces of news is the continued Japanese penetration of ever more sophisticated high-tech markets. According to L. William Krauze, chairman of the American Electronic Association, "the Japanese are eating their way up the electronic food chain." One of the critical areas is semiconductors, for which 50 percent of the world market is now dominated by Japanese chip makers. Japanese domination is total in the case of dynamic random-access memories, from the simplest to the most sophisticated. (During the last months of 1988, Toshiba was the first to produce 4-megabit DRAMs.) Japanese firms are also ahead in ASIC gate arrays and in optoelectronic circuits. They are making rapid progress in superconductivity, another area of intense competition between the United States and Japan. American firms are going to lose this battle, a report from the Defense Science Board warned in January 1989, unless the US research and development effort in superconductivity is increased by 50 percent. The race between Japan and the United States is also close in the field of supercomputers. In December 1988, Fujitsu announced a new family of machines using a 64-bit processor making 4 billion floating operations per second (4 gigaflops), which is the highest speed ever reached by a single processor.

Meanwhile, the Europeans have increased their pressure on several markets within the US sheltered culture. The announcement in 1987 of the new generation of Airbuses—the A-330 and the A-340—was received with great interest by some American airline companies. The launching of this new generation of advanced airplanes underlined the European consortium's ambition to compete with Boeing and McDonnell Douglas in all segments of the jetliner market. International Leasing Company, a California aircraft-leasing firm, ordered A-330s in July 1988. It was followed in January 1989 by Northwest Airlines, which ordered ten units of the new European airplane.

But the most important news from Europe since early 1988 has been coming from the exposed technical culture. Indeed, several recent events suggest that the Europeans are strengthening their capabilities in this domain too. One of the most spectacular events has been the creation of Jessi, a research consortium set up by Philips, Siemens, and SGS-Thomson for experimenting with the manufacturing technologies needed to produce the next generations of semiconductor memories. In many aspects Jessi is like Sematech, the Austin-based US semiconductor consortium: all major European industrial players in semiconductors have joined Jessi, and the project will also benefit from substantial government subsidies. But the most important point about Jessi is neither technical nor financial; it is that, for the first time in history, hereditary enemies in the European semiconductor industry have agreed to work together on tomorrow's technologies in order to contend with Japanese and American competitors. European companies have also undertaken major acquisitions and joint ventures in the United States, showing increasing aggressiveness in American exposed-culture markets. Among the most spectacular have been the acquisition of the RCA-GE consumer electronics business by Thomson and the joint venture headed by Bull with Honeywell and NEC.

High-definition television is another area where the Europeans seem to have made substantial progress. Until recently the Japanese NHK standard was the only one available. The standard will determine the development of a large array of new products—TV sets, VCRs, cameras, satellites, electronic components, and transmission systems. During the summer of 1986, four European manufacturers—Thomson, Bosch, Thorn-EMI, and Philips—agreed to cooperate, within the framework of the Eureka project, in developing a complete HDTV system, which was presented to the public for the first time in September 1988. This system is based on a set of standards to which all the European countries have agreed, including a transmission standard already being used experimentally on TDF1, the French high-power direct-broadcast TV satellite, which was placed into orbit in October 1988. The European HDTV standard has a significant advantage over its Japanese rival in that it is compatible with currently available TV receivers. For the moment the United States seems to be lagging in this new domain. Although technical capabilities exist to develop an American HDTV system, the main problem is the virtual absence of US television manufacturers. Indeed, in the face of foreign competition, America has progressively given up manufacturing consumer electronic goods. Most television sets sold today in the United States are manufactured by Japanese or European companies, which are therefore likely to be the major players in the new battle for HDTV markets.

There is nevertheless some good news for the United States in this landscape. The most important positive development has probably been the continuous—even though slower and more limited than expected—improvement in 1988 of the US balance of trade for industrial goods, and particularly for high-technology products. This trend, which began in 1986, is clearly the consequence of a lower exchange rate for the dollar against the chief competitors' currencies, particularly those of Japan and most European countries. A cheaper dollar—at times 50 percent lower than 3 years before—has boosted the competitiveness of American-made industrial products and has generated a new surge of exports. (US exports increased by 23 percent in 1988.) Indeed, in many industrial sectors American factories are humming again. So are US manufacturing facilities controlled by foreign companies, which are also taking advantage of a cheaper "Made in USA" label. For example, Honda is now exporting to Japan spare parts produced in its Ohio plant, thereby contributing to a reduction of Japan's trade surplus with the United States.

But fluctuations in the exchange rate have brought only temporary relief to US companies. In the long run they have no other choice but to improve their own productivity. In spite of a certain number of disappointments (GM is one example), factory automation continues to be one of the most obvious answers to this problem. In 1988 almost $20 billion was invested in this domain in the United States, but this is still lower than Japan's expenditure in terms of dollars invested per worker. Congress addressed this issue in its 1988 trade bill, making the improvement of productivity a major theme for government action. Among various proposals for boosting American productivity is the transformation of the National Bureau of Standards into the National Institute for Standards and Technology, a proposal that has already been implemented by the federal government. The NIST has been given a broad mission to foster technology transfers and the use of manufacturing technologies in industry, particularly in the case of small and medium-size firms. This new institute should play a key role in the development of regional centers for the transfer of manufacturing technology.

But the most important developments in this domain have resulted from initiatives taken by industry. Cooperative research between rival firms in a given industry has made a modest but promising start in the United States. Sematech, created in 1987 by twenty companies in industries that produce or use semiconductors and chips, illustrates this new form of partnership. More recently, leaders of the semiconductor industry have even seemed ready to go a step further in pooling their companies' resources against Japan. In January 1989, the Semiconductor Industry Association and the American Electronic Association announced the creation of a manufacturing consortium to produce 1-megabit DRAMs.

In the American sheltered culture, 1988 was a year of both relief and new uncertainties. The relief came from the resumption of the space shuttle program with the successful launch of *Discovery* in October. America is back in space after the long and painful interruption that followed the *Challenger* disaster. Moreover, two new expendable launch vehicles developed by the Air Force in the aftermath of the *Challenger* accident—the Martin Marietta Titan 4 and the McDonnell Douglas Delta 2—became operational in early 1989 and will be available for the launching of military and commercial satellites. The year 1988 also brought European and Japanese commitments to the US space station.

It would be premature, though, to declare that everything is back on track in the American space program, and that NASA should ready itself again for performance and glory. There is a growing consciousness in government circles and among experts that some choices will have to be made—primarily for budgetary reasons—among the various technology-oriented megaprojects, military and civilian, launched during the Reagan era. Moreover, in a time of scarce resources, these choices will necessarily come into conflict with yet another growing priority for government action: the strengthening of US industrial competitiveness.

For decades, the US government's involvement in technological development has been guided by the implicit principle that such involvement was appropriate for technologies related to public goods, such as defense and health, but was inappropriate for technologies aimed at the marketplace. This is why, historically, government programs for research and technology have led to the pouring of virtually unlimited resources—mainly through the Pentagon and NASA—into the sheltered culture, and practically no resources (except for basic science support) into the exposed culture.

This dichotomy was not a problem as long as companies in the exposed culture were also benefiting from military funds, as was clearly the case in the early years of the semiconductor and computer industries. But as time passed, fewer transfers occurred from the sheltered culture to the exposed culture. Moreover, the concentration of public resources in the sheltered culture soon started to create important distortions in the US high-tech system, contributing to the scarcity of skilled manpower—a critical parameter for companies exposed to foreign competition. Today, for the first time in history, the megaprojects in the sheltered culture are drawing on resources that could be crucial to industrial competitiveness in the exposed sector.

Events that took place in 1988 suggest that there is in the federal government a willingness to start correcting this. The Sematech project, jointly financed by industry and the Pentagon, seems to have become a new model for government intervention. In 1988 the federal government encouraged the creation of several other cooperative research ventures between companies in areas of intense foreign competition, such as superconductors, high-definition TV, and biotechnology.

During the same year, the debate broadened over the federal role in the development of commercially oriented technologies. Strengthening US industrial competitiveness has become a national objective for the government as well as for Congress. But it remains to be seen if this call for national mobilization will be effective enough. Setting ambitious goals for America's technology has always been, for the federal government, the most obvious way to mobilize resources, talent, and enthusiasm. Although 25 years have passed since the inception of the Apollo program, it remains a model for success in this domain. Even though the effectiveness of this approach has been seriously questioned (recall the unattainable goal of SDI and the *Challenger* tragedy), companies in the sheltered culture have always been more than willing to follow the Pentagon and NASA to technological achievement and glory. For the man in the street, meeting these technological challenges has become part of the American Dream.

But the challenge that America faces today is less prestigious and more difficult than going to the moon or creating a new generation of advanced weapons. It is also far beyond the grasp of the administration, since its major actors are the thousands of companies in the exposed culture fighting against foreign competitors. This does not mean, however, that it cannot be done. A year after writing the conclusion to this book I am still convinced that the American system is slowly adjusting to its new international environment. In doing so, the system itself may undergo profound changes. Grandiose endeavors will have to be traded for more modest day-to-day improvements. Less glory and more efficiency would seem to be the new fate of American high technology. As a friend of America, I hope that in this quest for competitiveness and efficiency there will still be room for the Dream.

Paris, March 1989

The New Frontier

1

When Ronald Reagan came to power in 1981, a long desert crossing was just ending for the American space agency. The last Apollo mission had been launched in 1972. Since then, the National Aeronautics and Space Administration had been grounded, its energies focused on the development of the space shuttle. By the beginning of 1981, work was almost complete. The shuttle *Columbia* was at last on its pad, waiting for a "go" from the new president.

The inaugural launch was scheduled for April 12. For the first time since the end of the Apollo program, the media were en route to Cape Kennedy to cover what promised to be a historic space event: the launch of a reusable vehicle that would mean a revolution in the technology of reaching orbit. On this occasion NASA was also to accomplish another technological first: the craft had never been tested in space.

On the morning of April 12 the long fueling procedure begins, preparing *Columbia* for launch. 145,000 gallons of liquid oxygen are pumped into the upper tanks, 385,000 gallons of liquid hydrogen into the lower. The crew has already been aboard for two hours. At T minus 6 minutes, the command center's computer takes control of a complex sequence involving several thousand last-minute checks. At T minus 30 seconds, four onboard computers take over. The shuttle's engines ignite at T minus 6.8 seconds. For an instant the craft stands in unstable equilibrium; then the ignition of the two solid-fuel boosters lifts its 2,045 tons from the earth. At T plus 9 minutes and 45 seconds, with astonishing precision, *Columbia* reaches its assigned orbit. Thanks to onboard TV cameras, the world can watch the astronauts floating about the cabin.[1]

After slightly more than two days in space, John Young, the mission commander, begins the reentry procedures. On April 14 at 1:21 P.M., *Columbia* touches down without problems. The shuttle's first flight has gone as well as NASA technicians could have hoped.

From Washington, Ronald Reagan pays homage to the "new pioneers of modern times." The American public is jubilant about the men whom NASA is already calling "space workers."

The American space agency has just reaped the fruit of great patience. A new era in the conquest of space seems to be beginning. It is the result of a difficult choice made ten years earlier by Tom Paine, then the administrator of NASA. For him the future of American space flight was centered on a single problem: As the Apollo program was ending, how could NASA take advantage of the American public's enthusiasm for space, sparked by the lunar landings, and convince President Richard Nixon of the need to develop a new manned space program?

Since his inauguration, in 1969, Nixon had never hidden his skepticism about the space program or his feelings that NASA's budget should be reduced.[2]

Ironically, it was at the very point in its history when dividends could be expected from the mission assigned to it ten years earlier by John Kennedy that the space agency was suddenly to find it hardest to safeguard its future. The winds of history had veered significantly. From the praise of scientific and technological progress underlying numerous official speeches during the Kennedy and Johnson administrations, there followed during the 1970s an era of doubt. The Vietnam war, the growing influence of the ecology movement, and questions about the model for Western economic development ("Is it time to halt growth?" wondered the Club of Rome) all played a role in the new disenchantment with science and technical progress.[3]

In this setting it seemed impossible to Tom Paine that he could hope to convince the president to commit the nation to an ambitious new project, a logical sequel to Apollo such as building a base on the moon or planning a trip to Mars.

Within the Nixon administration the mood was antithetical to such ventures. If it wished to receive public money, NASA should henceforth become accountable for its decisions and undertake truly useful programs for the country. In view of the growing demand for communications and observation satellites, it should devote its energy to developing the technologies they required. But it was impossible for NASA to forgo the very thing the Apollo program had demonstrated: the usefulness of men in space.

It was from such contradictory demands that the space shuttle was born. It would make manned space flight a commonplace thing, and greatly reduce the cost of launching craft into orbit. The shuttle would be a multipurpose vehicle, capable of carrying both astronauts and large payloads. With the emerging development of commercial space applications, the shuttle would make it possible to launch telecommunications satellites at the lowest cost, and it would satisfy the needs of both the Pentagon and the scientific community while providing the capability for putting humans into orbit to repair satellites and to assemble to the scientist-manned space stations of the future.

It was to the first step in this grand scenario that President Reagan was paying tribute as the *Columbia* crew returned to earth on April 14, 1981.[4]

New Heights for High Tech

Now, as at the time of the first manned lunar landing, a recently elected Republican president was reaping political dividends from an accomplishment with which he had had nothing to do.[5]

Discounting the events of the day, circumstances appeared no more favorable for a major undertaking in space than in 1969. Traditionally, by virtue of their instinctive aversion to "big government,"[6] Republicans have generally sought to clip the wings of government programs. Their avowed hostility to government intervention in the country's economic affairs, in the name of free-market principles, never allied them to major technological endeavors. The new president was apparently no exception in this regard.

In November 1980, Ronald Reagan was brought to power by the conservative wave that seems always latent in the United States during times of crisis. But the crisis of the late 1970s was not solely economic. As Jimmy Carter's term expired, America seemed to have lost confidence in itself, and seemed to be looking for a leader. With his indisputable media talent, the former governor of California was rapidly perceived as the "man for the job," in contrast to an outgoing president who had been discredited and weakened by the hostage affair in Iran. The ideas he expressed during his campaign were in no way original, but they had the merit of being simple and of meeting the expectations of a large part of the electorate. Ronald Reagan was calling for a return to basics, to the traditional values that had made America strong. Free enterprise, a taste for risk, and an ambition for success were in need of rehabilitation. As to the federal bureaucracy, usurping a growing part of the national wealth—there in itself was a major culprit for what the country was suffering. Lower taxes, reduce federal meddling to a strict minimum, and bring prosperity back to America: this was candidate Reagan's grand plan.[7]

On Wall Street, Ronald Reagan's election was greeted with satisfaction. A newcomer to the White House who was an apostle of free enterprise and a free-market economy could only be good for business. The stock market's reaction was immediate. During the first six months of 1981, the Standard & Poor's 500 index, one of the major indicators of the American economy, advanced by 20 percent.[8]

In Silicon Valley, events in Washington are usually watched with only a distracted glance. However, the former California governor's rise to the presidency made this election particularly significant. Not that Ronald Reagan was part of the local culture. By all expectations, the former actor from Southern California[9] and the high-tech raider elite around San Francisco could not have been farther apart; however, he talked their language and reflected their preoccupations.

Located 50 miles south of San Francisco, Silicon Valley is one of the temples of American high technology, a region where semiconductor and computer manufacturers grow like mushrooms. It is a promised land for the adventurers of modern capitalism, who bet on high tech in very much the same way that gamblers bet on cards or roulette.[10]

The majority of the Silicon Valley's business leaders are conservative Republicans. Free enterprise, personal initiative, and the law of the market are the values that carried them to success. They succeeded as entrepreneurs, and so they are convinced that the system in which they excelled is profoundly just and good. For them, to hold any other opinion would be to cast doubt on their own success. In Silicon Valley, lack of money or lack of strength can only reflect an incapacity for taking advantage of a society that offers endless opportunities to those who know how to take advantage of them. "These industrial leaders' belief in the virtues of a free-market economy," writes Stanford professor Everett Rogers, "go so far as to be absurd."[11]

Silicon Valley entrepreneurs rarely become involved in politics. Their compulsive individualism and the conviction that they themselves hold all the cards to their future keep them out of that arena. Exceptional circumstances are needed for one of them to cross the Rubicon.

Nevertheless, such things do occur. In 1982, Ed Zschau, director of a computer disk manufacturing firm, was elected to Congress from California's twelfth district, which includes Silicon Valley. Zschau discovered politics through his business. In 1978 he presided over a new committee set up by the American Electronics Association. The objective of this committee was to call the attention of the White House and of Congress to the need to lower capital-gains taxes, since the 49 percent then applicable was dissuasive for venture-capital investment. Initiatives taken by Zschau finally paid off. In the spring of 1979, Congress voted to reduce the capital-gains tax to 28 percent. This was to pave the way for unprecedented expansion of venture-capital investments, and of Ed Zschau's political fortunes.

The day after the election, on November 3, 1982, Ed Zschau addressed his peers, the Silicon Valley entrepreneurs, at a conference on innovation held in Palo Alto. "The role of government in encouraging innovation," he said, "is to create an environment with the freedom to succeed or fail in business. Washington must provide incentives, like a favorable tax structure." Zschau was loudly applauded. "This statement expressed Silicon Valley values," says Everett Rogers.[12]

This language was no different from that used by Ronald Reagan two months later before the Massachusetts High Technology Council, after a visit to electronics firms along Route 128. To the entrepreneurs of this other temple of American high technology, on the outskirts of Boston, Reagan said he wanted to be "the apostle of their success." A few days later, in his annual State of the Union speech,[13] Reagan expressed for the first time his vision of the role of technology. More than anything else, for him it was a source of amazement: "To many of us now, computers, silicon chips, data processing, cybernetics, and all the other innovations of the dawning high technology are as mystifying as the combustion engine must have been when that first Model T rattled down Main Street, USA." But these objects that so surprised us also pointed out a new source of wealth, a new frontier for exploration and conquest. The President continued: "But just as surely as America's pioneer spirit made us the industrial giant of the twentieth century, the same pioneer spirit today is opening up on another vast frontier of opportunity, the frontier of high technolgy." Ronald Reagan's vision of technical progress, inspired by the living example of the entrepreneurs he had visited along Route 128 and in Silicon Valley,[14] is perfectly consistent with the world view of the most conservative wing of the Republican party. For these advocates of free enterprise and a free-market economy, technology was indeed the new land of opportunity alluded to by Ronald Reagan. Technology could indefinitely extend the limits of what was possible. In a sense, it was the new fuel running the machine.

A Grand Presidential Scheme

If technology was a major concern for business, President Reagan soon discovered that it was also one for the military. Faced with the Soviet Union, America could recover its power and influence in the world only by reinforcing its defense, said the new president. A show of force is the only way, he explained, to start fruitful discussions with Moscow.

The first consequences of this commitment were to appear in the Pentagon's budget. In 1981 it increased by 18 percent. From $127 billion in 1980, by 1985 it had reached $285 billion[15]—the highest level, adjusted for inflation, in 30 years. With more than 7 percent of its Gross National Product (GNP) allocated for military expenses, the United States was now devoting as much of its resources to defense as it had during the war in Vietnam.

To begin with, Reagan insisted, the United States needed to recover the advantage in strategic arms with respect to the Soviet Union that had been lost over the years, particularly in Europe. Making attack impossible by raising the retaliatory cost too high was the logic of deterrence, the strategic concept whose application had for 40 years maintained a precarious balance between East and West[16]—a logic that had only fueled the arms race and encouraged the continual development of weapons.

In asking Congress for an unprecedented increase in research and development funds for defense, the new president was only reinforcing this way of thinking.

From the stealth bomber, whose materials and profile would leave hardly a trace on enemy radar screens, to the new generation of MX missiles, the drawing boards were cleared for new equipment and weapons systems. If the essential thrust was in offensive-weapons research, Reagan's arrival also brought renewed interest in anti-missile systems. Begun in the 1950s, research in this field had led to a study of several types of nuclear weapons. The concept envisioned at the time—proposed by Edward Teller, the father of the hydrogen bomb—involved detonating nuclear charges at high altitudes in the path of incoming enemy missiles. Richard Nixon gave the project approval in 1970 by authorizing experimental deployment in North Dakota of the Safeguard anti-missile system. The decision was highly controversial, not only because of the many technical problems involved but also because the opinions of most experts had by then changed significantly. Since the development of intercontinental ballistic missiles carrying multiple nuclear warheads, most specialists had agreed that the creation of an effective protective screen would be extremely difficult if not impossible, and that it would destabilize the balance of power. In any event, the signing in 1972 of the Anti-Ballistic Missile (ABM) Treaty brought the Safeguard project to a close.[17]

Edward Teller nevertheless remained an unconditional advocate of anti-missile weapons. In 1973, while he was still director of the Lawrence Livermore Laboratory, one of the two great American nuclear weapons research laboratories that he had created 20 years earlier, Teller assigned Lowell Wood to manage the "O Group," with the task of studying new nuclear weapons systems. Wood, a brilliant astrophysicist from the University of California at Los Angeles, became Teller's disciple and protégé.[18] He soon was concentrating the work of part of his group on the development of an x-ray laser.

Since their invention in 1960, lasers had held great fascination for the physicists involved in weapons development. Unlike the light from an incandescent bulb, which covers a wide spectrum of frequencies, the energy in a laser is concentrated in a narrow electromagnetic band. Furthermore, all the light "particles," or photons, are in phase. This means that large amounts of energy can be transmitted in a very short time, via a narrow beam that preserves its power over very great distances.

Lasers were soon to find multiple applications, from eye surgery to micro-electronics. But for the scientists working on defense systems their importance went even further. Now at last it was becoming possible to make an old dream of science fiction come true: the death ray. One needed "only" to build a more powerful laser generating a much shorter wavelength, in the x-ray band, and the job would be done.

The members of the O Group were not the first to believe that the energy from a nuclear explosion was probably the only kind that would be sufficient to trigger an x-ray laser. The job of designing the proper configuration between the "match" and the amplifying cavity remained.

On November 14, 1980, Project Dolphin was carried out successfully in the Nevada desert.[19] For the first time, it seems (for the test results were controversial), energy from an underground nuclear blast had produced an x-ray laser effect on a wavelength of only 14 angstroms.[20]

Edward Teller's dream had suddenly become a reality. The existence of a laser weapon was to shake up defense thinking. A single nuclear charge was enough to set off dozens of lasers aimed at enemy missiles. Rapidly, Teller developed the "pop-up" concept, illustrating his vision of strategic defense, made possible by the work of the O Group. The device producing the x-ray laser effect—the nuclear charge, laser tubes, sensors, and control electronics—

could be placed aboard a missile and launched into orbit the moment an alert was sounded. The sensors would aim the laser tubes at the enemy missiles, and then the nuclear charge would be exploded. The number of launches could be increased as demanded by the threat.

In February 1981, *Aviation Week and Space Technology* spoke of the Nevada test as a technological breakthrough having "the potential to stop a Soviet attack."[21]

During the summer of 1979, candidate Reagan visited NORAD headquarters, the Colorado nerve center of the American nuclear deterrent. He was highly impressed, it seems, by the sophistication of the military arsenal, and its complexity, but was surprised to see that "in spite of all this technology, we still can't stop Russian missiles coming down on us." During the election campaign he asked two of his advisers whether a theme of defense against nuclear weapons would not appeal to the electorate more than the theme of accelerated offensive weapons.[22]

Beginning in the summer of 1981, a small group of scientists, manufacturers, and high-ranking officers began meeting at the Heritage Foundation, a conservative Washington "think tank."[23] Their objective was to formulate proposals for a national defense system to be submitted to the president. The group included Edward Teller, along with several of Ronald Reagan's personal advisers, such as Joseph Coors and former Deputy Secretary of Defense Karl Bendetsen. The group met with the president for the first time in January 1982, and subsequently made two other visits to the White House. Teller himself had a private meeting with the president in September of that year, during which he was able to emphasize the importance of the technologies developed at Livermore. In all, Teller met with the President four times in less than a year.[24]

On March 23, 1983, in a solemn nationwide television speech, President Reagan set forth his vision for the defense of the United States and the free world. He used Teller's ideas as inspiration, but he went much further. The president proposed a new program, the Strategic Defense Initiative, whose creation would herald a new era for the world's security. In sharing his vision with the nation he said: "But what if free people could live secure in the knowledge that their security did not rest on the threat of instant US retaliation to deter a Soviet attack; that we could intercept and destroy strategic

ballistic missiles before they reached our soil or that of our allies?
. . . I am calling on the scientists that gave us nuclear weapons to
use their enormous talents in the service of humanity and world
peace; to give us the means of making those nuclear weapons
impotent and outdated."

The creation of an anti-missile system in space would make
this dream come true. Such a system could already be envisioned
on the basis of recent scientific work. It would require a great deal
of additional research, however. The government of the United
States had decided to spend $29 billion on this effort over a five-
year period in order to explore the different technical options and
study the feasibility of the project. A decision as to whether to
deploy the new defense system would be made at the end of this
period, before the end of the decade.[25]

The "Star Wars" project, as some opponents and the media
dubbed it, made the front pages at once. It set off a violent controversy, not only in the small circle of strategic armaments specialists
but throughout the scientific community. In Europe, "Star Wars"
produced some concern. What would its impact be on the defense
of the Old World? In view of the large sums involved in research,
would this program not produce an even greater technological gap
between the United States and Europe?

The American industries and media specializing in armaments
and in space were enthusiastic. Even if the underlying concepts
were open to question, what was important was that a new era was
opening for research and space activities. In short order a new
administrative body, the SDI Office, was set up under the Department of Defense, headed by General James Abrahamson. Its purpose was to coordinate the project and organize calls for proposals
from manufacturers and university laboratories, which would be
reaping the SDI harvest. As of 1984, $500 million of the Pentagon's
budget was allocated to SDI. Beginning in the fall of 1983, a large
number of space and defense journals were publishing editorials
and articles about the new promise of space and wondering about
the strategic, technological, and industrial consequences of the president's decision. In the space and defense industries there was
euphoria. In addition to the $29 billion to be invested just in the
research phase, at least another $100 billion would ultimately be

spent—the lowest likely cost of building and deploying the defense system.

For manufacturers in the military-space complex, SDI had been perfectly timed to create new needs. It came along at a point when most orders relating to the second-generation nuclear arsenal had already been filled.

At NASA, there were questions in the air. There had always been a fairly strict separation between civilian and military space programs. Would the new priority for the latter, created by SDI, work to the detriment of the civilian program? Relations between the Air Force and NASA had deteriorated in recent months. Delays in launching the space shuttle could endanger national security—this was an argument developed by the Air Force to reopen a subject that NASA had managed, with difficulty, to close some years earlier: the issue of expendable launch vehicles (that is, traditional rockets). For NASA, one condition for the success of the shuttle program was its ability to fulfill all launch requirements, both civilian and military. It was only reluctantly that the Air Force agreed to this position, abandoning efforts to upgrade its own launch capabilities.[26]

Another problem for NASA was the attitude of David Stockman, whom Reagan had named director of the Office of Management and Budget in 1981. Stockman was asking NASA to provide proof of its profitability before he would maintain the financing for the shuttle program. He had taken out the 1973 cost figures on which Richard Nixon had based his decision. At the time, the figure reluctantly provided by James Fletcher, then NASA administrator, was $100 per pound placed in orbit.[27] Now, James Beggs, who had just taken NASA's destiny into his hands, knew better than anyone else that the true cost for a pound in orbit was several thousand dollars—a cost that was not, after all, surprising. The shuttle, the most complex space machine ever built, was still only in an experimental stage of development. But it was out of the question for NASA to admit it. Under pressure from the OMB, from Congress, and from the White House, Beggs decided to take a few shortcuts. The shuttle's development phase would be limited to four launches, to enable it to carry commercial satellites as soon as possible.

However, for Beggs the most important gamble for NASA's future was the project that aimed to place a permanent manned station in orbit. That project had been submitted for presidential approval several months earlier. For NASA, the space station constituted an indispensable sequel and the logical next step after the shuttle program. Its total cost would be on the order of $6 billion. Permanently inhabited by teams of scientists, it would serve as a base for earth observation and for experiments in the gravity-free manufacture of chemicals and drugs. It would also be a platform for astronomical observations.

The real reason for the space station, however, was political. While awaiting authorization to take on more ambitious missions, such as planetary exploration, NASA urgently needed to receive assignments that would enable it to capitalize, with respect to technology and to public opinion, on its experience acquired in manned space flight. The orbiting station could fulfill both requirements. Furthermore, it was the only major operational project that James Beggs found waiting when he took on his job at NASA in 1981.

The project was far from being unanimously approved, however. Scientists were complaining about the excessive priority accorded by the agency to manned space flight, to the detriment of automated experiments (which were less expensive but just as valuable from a scientific viewpoint, as the Voyager 2 probe had brilliantly demonstrated by sending superb pictures of Saturn from a distance of several million miles in August 1981). In 1984 the Office of Technology Assessment, Congress's technological arm, published an extremely critical report on the proposed space station.[28] But it was too late to stop a project that was already well advanced in the decision-making process. The Committee for Science and Technology of the House of Representatives allowed itself to be convinced by James Beggs. So did the Senate committee in charge of the space program's budget. President Reagan had made his decision several months earlier. It was positive. He had made it in spite of the avowed hostility toward the project of his adviser for science and technology, George Keyworth. A confirmed partisan and advocate of SDI, this former physicist from the atomic weapons research laboratories at Los Alamos[29] was also close to Edward Teller, who in 1981 had introduced him to the president. For him the space station was nothing but "a hotel in the sky for astronauts."[30] What was required, in his mind, was a grand scheme for space that would meet with the public's approval and enthusiasm, just as the Apollo program had.

NASA's objective, thought Keyworth, should be either a manned base on the moon or a journey to Mars. According to him, such an ambitious vision of the role of space technology was shared by the president. He had tried—in vain, he said—to convince Beggs. Here is how Keyworth recalled the meeting he had arranged in late 1981 between the NASA administrator and President Reagan: ". . . after a few remarks about progress on the shuttle program, the president asked James Beggs what NASA was planning for the future. He answered with extensive details about the space station. After finishing, and while he was waiting for a sign of interest if not of approval, the president simply said, in a tone that obviously reflected his disappointment, 'is that all?'[31]

This disappointment did not prevent Ronald Reagan from giving his support to NASA, which, a few months later, got the green light to start preliminary studies for the space station. He also approved the space agency's idea of putting nonprofessionals aboard the shuttle. After a brief time in orbit, he believed, members of Congress would be the first to testify that space technology had come of age, making the earth's suburbs accessible to all.

A Hymn to Technology

Ronald Reagan's confidence in high technology went far beyond what it represented as an opportunity for business. In fact, technology itself seems to have been a source of fascination for the new resident of the White House. From "Star Wars" to the space station, via Route 128 and Silicon Valley, the new president's confidence in science and technical progress seemed unshakable.

Such faith in technology was nothing new for a president of the United States. Its foundations can be found in a report that Franklin D. Roosevelt requested of Vannevar Bush in November 1944. Entitled "Science: The Endless Frontier,"[32] it presented a broad panorama of the benefits that could be expected from technological progress during the postwar era. Eighteen years later John F. Kennedy was still inspired by it. The importance he attached to technological development was reflected not only in the goal he set for the space program—a man on the moon before the end of the decade—but also in his avowed conviction that science should be able to provide the answers to a great number of problems facing mankind. Such a belief also lay behind Lyndon Johnson's "Great Society." It persisted as well through the first term of Richard Nixon, who initiated a cancer-research program of unprecedented magnitude.[33] According to Harvard professor Harvey Brooks, a prominent specialist in American science policy, even during the period of disenchantment that followed, faith in technology never faded at the highest levels of the US government.[34] This faith became an integral part of the Reagan administration's political commitment. Technology was the new frontier of free enterprise, and it would provide an answer to the most formidable problem ever to face humanity: halting the nuclear arms race.

The president's SDI speech favorably impressed financial circles on Wall Street. Stock quotations for McDonnell Douglas and Martin Marietta, two of the Pentagon's traditional contractors that stood to benefit enormously from SDI contracts, rose by 20 percent during the last six months of 1983. In fact, for a year and half already, the stock market had seemed bewitched by high technology. The High Technology Stock Growth Index, which represents the average value of a sample of small computer companies, had risen from 100 to 250 between September 1982 and May 1983.[35] From data processing to biotechnology by way of integrated circuits and telecommunications, investors everywhere felt the same enthusiasm for "high tech." The Hambrech & Quist Index, one of the indicators reflecting average changes in high-technology stocks, rose from 115 in July 1982 to a record 280 in May 1983.[36] For those who were gambling on the rising stars of Silicon Valley, the results had exceeded their most optimistic forecasts. For investors who had followed the advice of the venture capitalists, the financial gurus of Route 128 and Silicon Valley, high technology looked like a new Eldorado. The growth in the volume of funds collected by venture capitalists provides a significant index of the course of events: in 1980 it was $400 million; it reached $1.2 billion in 1981, $3 billion in 1982, and $4.5 billion in 1983.[37]

Already the Reagan vision of high technology seemed to be making its mark on history.

For a Few Million Dollars

2

At first glance there is nothing special about the small office buildings hidden among the trees at 3000 Sand Hill Road in Menlo Park,
California. The narrow access road winding among the hills of this
residential area near Stanford University has nothing distinctive
about it either. The only notable things on this road are the names
over the doors of various buidings, which remind the visitor that
this is the land of high technology: IBM, Lockheed, and Software
Inc., among others, are located here. The activities of the residents
of 3000 Sand Hill Road are better understood from the directory.
Among the hundred or so names listed are those of the most
renowned practitioners in a calling whose development began in
the 1960s around San Francisco: venture capitalist. Half a dozen
one-story buildings house the world's highest concentration of this
special breed of financier, whose main activity is to identify rising
stars in the firmament of high technology and provide them with
capital on which to grow. Sand Hill Road is the Wall Street of high
technology.[1]

Eugene Kleiner and Thomas Perkins are typical representatives
of the venture-capitalist profession. They established themselves in
Menlo Park in 1972. During their first five years in business they
invested $7 million in seventeen companies. Their first undertakings
turned out to be disastrous. By 1984 seven of the firms they had
bought into had gone bankrupt or were showing values far below
the amount of their initial investment. But two of the remaining
companies had enjoyed monumental increases.

One of them was Tandem Computers, whose founder, James Treybig, had been the first to tackle the problem of computer breakdowns. Treybig designed a program that enabled two computers to work in parallel, so that one could take over if the other failed. All the airline reservations systems were soon equipped with the Tandem system. Kleiner and Perkins's investment share of this company was now worth $152 million, for an initial stake of $1.45 million. [2]

The second company was Genetech, one of the first biotechnology companies. Kleiner and Perkins had invested $200,000 at a time when biotechnology looked like promising but risky. Five years later a large number of the biotechnology firms born in the late 1970s had disappeared. But Genetech had turned into one of the most outstanding successes in this new high-tech sector, and the value of Kleiner and Perkins's stock was $47 million. In all, their portfolio had gone in eight years from $7 million to $218 million.

The excellent profitability of this venture-capital fund soon enabled Kleiner and Perkins to raise capital for a second fund of $15 million, then a third of $55 million, and in 1982 a fourth of $150 million. By the end of 1984 they had invested a total of $133 million. Today the companies whose creation or development they financed are worth nearly $5 billion.[3]

Venture capitalism is a risky business, but it is also an art: the art of detecting, among the multitude of entrepreneurial candidates, who will be the Steve Jobs or the Robert Noyce of the future. The soundness of the projects from technical, commercial, and financial viewpoints is of course an essential criterion, but the venture capitalist's attention must be focused chiefly on the men who put them together. Before deciding, he wants to know their strengths, weaknesses, and ambitions. When he finally decides to invest in a company, the venture capitalist will propose nothing less than marriage. He wants to travel alongside the entrepreneur through the first years, which are often difficult and filled with setbacks, to provide advice, financial expertise, and new financial partners if they become necessary.

Consequently, the activity of a venture capitalist has little in common with that of a classic banker. Once assured that his customer is solvent, the banker has no intention of becoming involved in the company. It is not in his interest to do so. Lending money is a service provided at a fixed rate of interest, whatever happens. The only thing that counts for the banker is that the borrower be able to pay back his loan. Should he be unable to do so, it would mean a heavy loss. Since the banker's profit on loans is rarely greater than 2 percent, he would have to arrange fifty to a hundred similar operations to cover his loss. This is why bankers have such an aversion to risk. Most of the time they must say no to start-up entrepreneurs asking for loans. They simply cannot lend money to firms that do not even exist yet.

In contrast to bankers, the venture capitalist does not lend money; he buys an equity share in the company. If the company fails, he loses his investment; if it succeeds, he may recover his money many times over. A single successful operation may compensate for several failures. Profit for the venture capitalist will materialize down the road in the form of a capital gain, obtained by selling stock in the company once its market value has substantially increased. In fact, the heart of the business lies in helping young companies in which one has invested to grow and prosper until they can be introduced on the stock market, and then selling one's equity in order to start over.

The funds invested as venture capital generally come from institutional or private investors who are attracted by the high rate of potential profit. Here, tax policy plays an essential role. The decreases in capital-gains tax rates that occurred in 1979 and 1982 explain to a large extent the flow of money into venture-capital funds that was seen in the early 1980s.[4]

It was not by accident that Kleiner and Perkins chose the Bay Area in which to do business. Beginning in the late 1960s, a large number of new companies were appearing here in a new industrial sector: semiconductors. This field had two characteristics that venture capitalists seek: a rapidly developing technology offering numerous new opportunities for development, and fast-growing markets.[5]

It was William Shockley's invention of the field-effect transistor in 1951 that gave birth to this industry. Transistors were soon to replace large, fragile diodes and vacuum tubes, thus making it possible to process and store data under radically new conditions. New waves of innovation over the years diversified their applications and led to silicon "chips" that integrated a growing number of transistors and improved performance vastly. Their use gave birth to a no less remarkable flood of inventions. The availability of silicon chips not only produced a revolution in the information and communications industries; it also affected many traditional industries, such as household appliances and automobiles. Chips were soon being used in most industrial sectors.[6]

Two examples will suffice to illustrate the circumstances under which these developments occurred. They characterize a development mode, based on technology and innovation, that extends far beyond the Bay Area, where it was born. The first of these examples concerns the invention of the microprocessor; the second describes the upheaval produced in data processing by the advent of the microcomputer.

Of Micros and Men

When Intel announced in 1971 that it was marketing a silicon chip capable of performing alone all the data-processing functions of a computer, the profession greeted the news with a certain degree of skepticism. Caution was the order of the day in an industry where miraculous innovations were regularly being announced, the performance of which did not always live up to the claims. Furthermore, the idea of miniature computers excited no one. The tendency in the semiconductor industry was to use improved component performance to increase the speed and capacity of large computers, not to produce small ones, for which there appeared to be no demand.

The inventor of the microprocessor, a young Stanford engineering graduate named Ted Hoff, had joined Intel two years earlier. Upon mentioning his project to Robert Noyce, the head of Intel and one of its co-founders in 1968, Hoff received enthusiastic support. Two years later a revolution was set in motion when the company began to market its new product, the 4004 microprocessor.[7] Not only could this new chip replace a large number of logic circuits, but it would soon become the basic tool in the design of most electronic systems. Microprocessors could be used to build pocket calculators or to control computers, and could be used to "program" thousands of devices, from traffic lights to washing machines, elevators, and even industrial production lines.

Competitors reacted immediately. Four months after Intel's product reached the market, Texas Instruments announced a microprocessor with even better performance. Its arithmetic and logic circuits could handle 8-bit data words, whereas those of the Intel product could not handle words with more than 4 bits; thus it could process 16 times as much information per unit of time. By April 1972, Intel had improved its original product and was also offering an 8-bit chip—one with a lower price and higher performance than the Texas Instruments product. In August 1973, Intel introduced the 8080, the first 16-bit microprocessor, a second-generation device. Motorola, Altair, and National Semiconductor also began to produce microprocessors, the demand for which was growing dizzyingly. Intel managed to keep its lead thanks to intense research and development efforts. The 8080 was without doubt the best product of its generation. IBM chose it for use in its first personal computer—a decision that delighted Intel, which in 1985 controlled 85 percent of the market for 16-bit microprocessors. Since 1983 Intel's yearly sales had risen to over $1 billion.[8]

In February 1981, Intel announced the 432, the first 32-bit microprocessor. This made it possible to build a computer no larger than a telephone directory with the same computing power as a mid-range "mainframe." Intel also introduced the multiple processor, which was capable of handling several tasks simultaneously. Ahead of its time, the 432 would not be marketed until 1985, by which time Motorola, National Semiconductor, Fairchild, and AT&T had developed comparable products. The battle for control of the 32-bit-microprocessor market, with a yearly sales potential of 5 million units by 1990, was underway.[9]

If Intel's 1971 innovation marked a turning point in the semiconductor industry, it occurred in the context of a more fundamental evolution—one that is a singular case in the history of industrial development. Its prime characteristic was an acceleration in technological change that produced performance increases of several orders of magnitude. The reason for this improvement was a higher level of integration—an increase in the number of circuits, or memory cells, per unit of surface area. While they were becoming denser, chips were also becoming faster, since the distance that currents had to travel through circuits was reduced. This led to the "law" noted in 1964 by Gordon Moore, today the president of Intel, who was then working for Fairchild. Moore noticed that the number of circuits a chip could carry was doubling every year. He predicted that the phenomenon would persist because of the manufacturing experience that was being acquired and the room that remained for technological improvement.[10]

Along with this increased efficiency came sharply falling prices, leading to ever-increasing demand. The learning phenomenon, which exists in most manufacturing industries and which leads to cost reductions as the number of units produced increases, was in the case of semiconductors more significant than ever. The result was a cost decrease for each electronic function, so that the price of a silicon transistor fell by a factor of 100 between 1954 and 1972, and that of a digital intergrated circuit by a factor of 15 between 1964 and 1971.[11] In 1973 Intel's RAM memory capacity went from 1 to 4 kilobits—to 4,000 memory cells holding one bit of information each. In 1976 this capacity rose to 16K, then to 256K in 1979. During this period prices fell sharply as semiconductor manufacturers progressed along their learning curves, with the concomitant cost reductions as a function of the number of units manufactured.

Prices were also falling because of intense competition in the field, to which Japanese firms were contributing ever more actively. Sixty-four-kilobit RAM memories, which sold for $15 apiece in 1981, were going for $3 in 1983. The price of 256K memories fell from $110 in 1983 to less than $4 two years later.[12]

Only a few of the many firms that entered this fast-expanding market managed to stay in the race and to produce successive generations of products. Aside from the ever-increasing investments required for R&D as well as for production, another critical factor was time. In 1977 Robert Noyce noted that "a year's advantage in introducing a new product or new process can give a company a 25 percent cost advantage over competing companies."[13]

A chance for huge profits, but at high risk, was the rule of the semiconductor game.

Concentration within the industry remained practically constant from 1960 to 1980. During this period the four leading firms were sharing 40 percent of the market, the eight largest 60 percent. The top spots, however, often changed hands. Of the ten largest semiconductor manufacturers in 1960, only four remained in 1980: Texas Instruments, Motorola, Fairchild, and RCA. The turnover can be explained by the large number of newcomers to the field. During the first fifteen years of its existence, the silicon industry grew via a multitude of new companies, most of them spinoffs from nearby Stanford University or from existing firms. The classic example is Fairchild Semiconductors. Created in 1957 by eight former employees of the company Shockley had created a few years earlier, this firm gave birth to a large number of companies—among them some of the best-known names in the profession, such as National Semiconductors, Intel, and Advanced Micro Devices. At a semiconductor conference in Sunnyvale, California, in 1969, fewer than 25 of the 400 participants had never worked for Fairchild.[14]

This unprecented rate of spinoff[15] can be explained by the numerous opportunities offered by a technological development with an exceptional wealth of applications, as emphasized by Gordon Moore: "Any company active in the forefront of semiconductor technology uncovers far more opportunities than it is in a position to pursue. When people become enthusiastic about a particular new opportunity but are not allowed to pursue it, they become potential entrepreneurs. . . . when these potential entrepreneurs are backed by a plentiful source of venture capital, there is a burst of new enterprise."[16]

The role of the "alma mater"—whether a company or a university—is crucial in this process, for it is she that most often provides the new opportunity or idea. A list of the major innovations that paved the road toward industrial semiconductor development shows that most of them were born in large companies[17]:

• the transistor principle, discovered at Bell Laboratories in 1940
• integrated circuits, created by Texas Instruments and Fairchild in 1961
• the MOS transistor, invented at Fairchild in 1962
• the Gunn diode, invented at IBM in 1963
• the microprocessor, invented at Intel in 1971.

In reality, what happened during the first twenty years of the semiconductor industry is an example, since become classic, of complementarity between large and small firms in a growth dynamic.[18] Over the years this complementarity also grew for other reasons. In order to ensure their development, the firms that gave birth to the industry needed to attend increasingly to mass production, the only way to lower costs and thus remain competitive. Large productive investments were necessary, making it harder to get into the business. New companies nonetheless continued to play the essential role of ensuring transfer to the market of new ideas for products or services, and of implementing a large number of innovations for which the initial investment was less burdensome. This was true, for example, of the new sector that developed to produce custom-made integrated circuits. In contrast to standard circuits produced in large numbers, these could be manufactured in small series to meet users' specific requirements. The key to access to this market was the development of extremely complex software that could reduce the time needed to design circuits. Today this sector is experiencing intense competition among companies that are, for the most part, of recent creation.

The microcomputer industry emerged under different circumstances, but it has many features in common with the semiconductor industry. It provides another view of the subject, one that can help us better understand the characteristics of a model of development that can be extended to a good number of high-technology industries in the United States.

Apple, the departure point for the microcomputer industry, has long been part of Silicon Valley legend, along with Steve Jobs's garage in Los Altos, California, where it all began. With Steve Wozniak, a talented electronics hobbyist, Jobs began building microcomputers for his friends in his spare time. Soon he was offering to sell a few to the nearby computer store in Mountain View; the store took 25. This was in the summer of 1976. For some months, Jobs and Wozniak had been convinced that they had a viable product. They discussed it with Atari founder Nolan Bushnell, Jobs's boss, without success. They were also turned down at Hewlett-Packard, where Wozniak worked. The microcomputer market, evaluated at only a few million dollars in 1976, held no interest for such well-established firms, which were committed to very different products.

As a result, Jobs and Wozniak decided, in the autumn of 1976, to create their own company, with the help of Mark Markkula, former marketing manager at Intel. In a few weeks Markkula managed to collect $690,000 and a $250,000 credit line. Apple Computer, Inc., was born.[19] The company presented its first model, the Apple II, at the San Francisco computer fair in April 1977. By the end of the year Apple had sold $2.5 million worth of personal computers. The following year sales reached $15 million. They reached $117 million in 1980, and $538 million in 1983. By the end of 1983 the company had built and sold 750,000 Apple IIs. It had nearly 4,000 employees, and it was being quoted on the stock exchange. At age 28, its president, Steve Jobs, had a personal fortune worth an estimated $300 million.[20]

The example of Apple soon set off intense competition among manufacturers of personal computers. As fast as possible, capture a share of the wildly growing market (sales were doubling yearly)—this was the object of the game. By the end of 1982 there were no fewer than 40 microcomputer manufacturers in the American market, producing in all no less than $3.5 billion worth of equipment. Their production volume rose to $7.5 billion in 1983.[21] Tandy, Commodore, and Compaq were the new rising stars in microcomputing, and Apple's chief competitors. However, in 1981 the most redoubtable competitor of them all appeared: IBM. The computing giant's entry into the personal computer market suddenly aroused anxiety throughout this young industry.

IBM had followed Apple's success with interest. Its top management had been quick to recognize the significance of the innovation introduced by Steve Jobs. The decision was made in 1979: IBM would not be absent from this new market. But speed was of the essence if IBM was to acquire a share of this fast-changing field. Breaking with IBM tradition, general manager John Opel decided to assign to a small and totally independent group the task of coming out with a microcomputer in record-breaking time. Although designing a new model at IBM was generally a matter of years, William Lowe and a team of a dozen engineers and salesmen in Boca Raton took only 13 months to create, in 1981, the first IBM PC.[22] Aside from the fact that this new product was created essentially from components produced outside IBM, it was a departure from the company's tradition in one other respect: like the Apple, the IBM PC was designed to be an open data-processing system. Unlike traditional computers, for which software supplied by the manufacturer was a mandatory access bridge for all users, anybody who wanted to could write programs for the IBM PC. The strategy paid off. Data-processing service companies mobilized their energies to produce operating systems and applications software for the PC, and their availability turned out to be an essential advantage for stimulating sales. IBM handed the job of creating the first operating system for the PC to the young software company Microsoft. Ms-DOS was soon to become a standard in the profession.[23] IBM PC sales increased rapidly. By the end of 1982 the youngest member of the IBM family had won nearly 50 percent of the microcomputer market.

With the introduction of the PC, IBM was participating in a technological revolution that was already underway. In spite of its gigantic size—$35 billion in sales and more than 300,000 employees in 1982—the company had demonstrated a remarkable ability to adapt. IBM's strength comes from highly decentralized management, structured around product lines, which enables it to adapt to the demands of specialized markets—and also from a remarkably efficient sales division. "IBM salesmen have always had the reputation of knowing their customer's business better than he does himself," noted Andrew Pollack in the *New York Times*.[24]

This concern with anticipating customers' requirements led IBM to create one of the most efficient customer-service operations in existence. In any event, this is the opinion of Thomas Peters and Robert Waterman in the book *In Search of Excellence*, a discussion of the best American companies.[25] Installing a computer at a customer's place of business means not the end of a sale but the beginning of a new relationship. IBM was the first to understand this and to master the art of after-sales service, enabling its customers to benefit from the latest available improvements in hardware or operating software. This method proved to be remarkably effective for creating customer loyalty. During the 1960s, when IBM reigned supreme over data processing, the company imposed its own path of technical change on the entire industry. The appearance of successive generations of large computers only reinforced its dominant position. The situation changed, however, during the 1970s, with the diversification of data-processing equipment and services. This was the result of rapid technological progress that made it possible to lower the cost and increase the performance of computers to an enormous degree. The IBM PC had calculating power equal to 100 times the power of the standard computers being sold by the company in 1959. This capacity came from equipment that took up 50 times less space, at 50 times less cost. Such evolution created many opportunities for firms seeking to enter a data-processing market that was now highly diversified.[26] As the industry's structure changed, competition became more active. Although IBM lost ground in some markets in the early 1980s (it was then making only a third of all minicomputer sales), the company still exerted a dominant influence over the profession. The speed with which it was able to enter the microcomputer field is a sign of its remarkabale capacity to adapt.

A Development Model for High-Technology Industries

From this rapid overview of the circumstances that led to the development of the semiconductor and microcomputer industries we can identify some general characteristics for a model of innovation and industrial growth that will be applicable to a large number of high-technology sectors in the United States.

The first of these characteristics is the complementary role of large and small companies. The large firms provide a resource of know-how and, more than anything else, of manpower. New companies being spun off can implement ideas that were born and brought to maturity in the large ones. The new firms contribute to growth in the industry: 85 percent of the firms in Silicon Valley have fewer than 50 employees.[27]

This explains a great deal about the second characteristic of this pattern of industrial development: its geographic concentration. Nearly two-thirds of the semiconductor industry and nearly half of the computer industry are located in California and in the Northeast (New York and Massachusetts). Small high-technology firms cannot be born and grow in isolation. They need an environment that provides the resources for their development. The existence of a local environment with an abundance and a diversity of technical resources (from both universities and large companies), and with financial resources available from venture capitalists, is one of the essential conditions for the appearance and the growth of high-technology companies.

However, the most important key to this development is human resources. A number of engineers and budding entrepreneurs played crucial roles in the development of Silicon Valley. With the hindsight provided by history, some call them geniuses. Others feel that their chief virtue was to have been in the right place at the right time. Their merit remains intact, however, for they assumed the risks of developing their products and their companies.[28]

In general, they had multiple motives for undertaking such an adventure. A taste for risk and the possibility of making it in the marketplace motivate many budding entrepreneurs. The creation of one's own company is an unquestionable path to self-realization, and to recognition. Scientific or engineering curiosity may also play a part, at least at the beginning. Designing and manufacturing a machine that works is in itself an exciting enterprise. But once he has mastered the first stage of his innovation, the aspiring entrepreneur is faced with a problem that generally overshadows all others: money—the money he needs to start the business and the money he hopes to make if he succeeds. To become a millionaire is for most a cherished dream, if not an end in itself. This objective does not imply wealth alone. It is also synonymous with social success and prestige. In Silicon Valley a hierarchy of values exists that is measured by the yardstick of bank accounts. The dream that one's own account may someday contain a few million dollars is the principal engine driving the system.

The fortunes built overnight by such flamboyant entrepreneurs of high technology as Steve Jobs, Jimmy Treybig, and Jerry Sanders attracted many emulators. "High tech" represented the last promised land for financial and industrial adventure. As was observed by Everett Rogers: "The Silicon Valley game is played out by several thousand capitalist technologists involved in a myriad of deals, spinoffs, startups, successes, and failures. The role of the federal and [the] state government is close to zero. Unfettered market forces pass final judgment on the boom or bust of firms and of individuals. Silicon Valley is high technology capitalism run wild."[29]

The high-technology firms around Boston and San Francisco are certainly not representative of American industry as a whole, but they constitute its "leading edge" of competition and high performance if success in the market is to be used as a criterion.

This universe is truly the universe of the great industrial adventures of the New World. The Silicon Valley entrepreneur is a descendant of yesterday's great captains of industry. If Henry Ford or E. I. Du Pont were to reemerge among us today, his chosen field would undoubtedly be high technology.

This model is not exhaustive, however. Let us now explore an entirely different face of American high technology.

Under the Wing of the Pentagon

3

In 1982, thanks to $2.2 billion in contracts from the Pentagon and from NASA, the TRW Corporation became one of the giants of military and space electronics. Its volume of business in this sector had more than doubled over the previous five years and now represented 42 percent of turnover.

TRW was created in 1958 by two former Hughes engineers who had just obtained a contract for one of the first ICBM projects. A Cleveland manufacturer of valves and spare parts for airplanes provided facilities and served as a base for the new company. While maintaining the parent company's activity, the new firm rapidly developed its business in the most advanced sectors of military and space technology. TRW created the engine for the Apollo lunar landing module. It built the Pioneer II probe, as well as several satellites for Defense Department intelligence operations which specialists agreed were technologically advanced. TRW also developed a competence in microelectronics that, as of 1981, made it one of the companies best suited for participating in VHSIC, the new Defense Department research project in high-speed integrated circuits. In 1987 this research enabled the company to announce the forthcoming introduction of a "superchip"—a very-high-performance, very-large-scale-integrated electronic component with the equivalent of 35 million transistors on a single chip—for multiple applications in missile guidance and in other weapon systems.[1]

When, in 1984, requests for Strategic Defense Initiative proposals began to be issued, TRW had already acquired solid experience with lasers, with the most advanced telecommunications techniques, and with battle-management software. The company became a favored industrial partner of SDI Office, the administrative body responsible for distributing funds for the SDI program. With $570 million worth of contracts, by the end of 1986 TRW ranked third among the program's contractors, immediately following Lockheed and Hughes. TRW's position reflects the importance acquired by the electronics sector in the evolution of military technologies. The increased performance of silicon chips had generated an even greater revolution in the defense sector than in other industrial sectors. Guidance, control, detection, and alert systems, military communications, and aircraft flight-control systems were among the areas most affected by this microelectronics revolution.

For example, highly miniaturized computer-controlled guidance systems coupled to sensors providing highly precise descriptions of targets, based on data collected from satellites, gave birth to new generations of cruise missiles. Mostly it was progress in microelectronics that made it possible to transform this technology, invented during World War II, into a modern and devastatingly accurate weapon. The McDonnell Douglas TERCOM (Terrain Contour Matching) guidance system, which weighs only 37 grams, can guide a cruise missile to within a few dozen meters of its target. TERCOM uses an on-board computer that compares the topography of the terrain, measured by a radar altimeter, against the characteristics of the programmed trajectory, which it automatically corrects.[2]

In 1986 the Electronics Industry Association estimated that close to two-thirds of SDI research contracts involved the electronics sector. This research was broken down into five subprograms, whose names give an idea of the project's major priorities.[3]

The first of these subprograms, called SATKA (Surveillance, Acquisition, Tracking, and Kill Assessment), was focused on the study of technologies that can provide "eyes" for the surveillance, identification, tracking, and destruction of missiles.

The second, DEW, was aimed at developing different types of directed-energy weapons (e.g., lasers and particle beams).

The third, KEW (Kinetic Energy Weapons), dealt with the development of weapons using the kinetic energy of accelerated projectiles.

The fourth, SC&BM (Systems Concepts and Battle Management), was intended to design an "electronic brain" to process and coordinate the millions of data bits required to position and fire the defense system.

The fifth and last is SL&KT (Survivability, Lethality, and Key Technologies), whose purpose was to analyze the survival of equipment in a hostile environment and study its vulnerability to countermeasures.

By the end of 1987, the SDI Office had distributed $8 billion worth of research contracts to the armaments, space, and military electronics sectors.

A Technological Race

The SDI program is only the most recent and most complex stage in the long race of military technology. A brief historical review will help us better understand the setting and characteristics of this race.

World War II marked an important point in this evolution, because of the sudden change of dimension brought about by atomic weapons. The explosion of the first Soviet atomic bomb, in September 1949, was the starting signal for the arms race. In order to guarantee its security, the United States decided to maintain its lead by developing ever-improved nuclear weapons. The result was an unprecedented extension and diversification of the technological areas calling for development. For example, there was a need to miniaturize and increase the power of nuclear warheads and to study different means of carrying them into enemy territory. This development would lead to Polaris submarines, Titan and Minuteman missiles, B-47 and B-52 strategic bombers, and a vast repertoire of techniques for identification, guidance, and communication.[4]

Technological progress brought particularly spectacular gains in the area of aeronautics. Jet engines, based on a concept developed by the English firm Whittle, appeared on German Messerschmitt fighters near the end of World War II. They were also being developed in the United States around that time by General Electric and by Pratt & Whitney. They led to a new generation of aircraft whose performance greatly surpassed anything possible with propeller-driven planes. The Bell X-1 broke the sound barrier for the first time in 1947. From the McDonnell Douglas F-4 to Lockheed's F-11, combat fighters evolved toward ever-increased performance and complexity.

In parallel, Boeing developed successive generations of giant bombers for the Strategic Air Command, the nexus of America's dissuasive power through the early 1960s. The B-52 replaced the B-47 in the early 1950s, but attempts to develop the next generation of supersonic strategic bombers led to extremely expensive prototypes (the B-58 Hustler and the B-70 Valkyrie) that were never mass produced. Missiles began to replace airplanes for carrying nuclear warheads. However, a supersonic-bomber project re-emerged during the 1970s in the form of Rockwell's highly controversial B-1. This was the most expensive airplane ever envisioned. Loaded with electronics, it was equipped with sophisticated sensors and automatic guidance systems enabling it to fly at extremely low altitudes to avoid detection by enemy radar. It was designed to cruise no more than 200 feet above the ground at nearly the speed of sound. A decision to build the B-1, often postponed because of numerous technical problems, was eventually made by Ronald Reagan in 1982. Six years later, a few months before the scheduled delivery from the Rockwell plant in Los Angeles of the last units of this highly sophisticated strategic bomber, the total budget for the B-1 had already reached $27 billion for the hundred airplanes built.[5]

Development of the next generation of bombers, the B-2, was contracted to Northrop in 1981. In 1987 this program alone accounted for nearly half the company's turnover, and involved a third of its 45,000 employees. The B-2 is one of the first stealth aircraft, designed to be practically invisible to radar. Stealth research, which has been underway for many years, has developed aircraft with weak radar "signatures," obtained by using radar-absorbing composite materials and a low-reflection profile.[6] With stealth technology coming to maturity, the myth of an invisible airplane seems finally to be becoming a reality. The high stakes explain the extreme secrecy surrounding the program—a program that is expected to cost some $50 billion by the time 132 B-2s replace the B-1s in the late 1990s. In January 1988 the Department of Defense gave Northrop the go-ahead to begin production, along with a $2 billion down payment.[7]

Space is another area where technological change has been very rapid. The Soviet Union's launching of Sputnik, in 1957, shifted the emphasis of the Pentagon's technological programs to this sector.

The space race had in fact been underway since the early 1950s, when the military became conscious of the potential importance of observation and spy satellites. However, the first American satellite was nonmilitary. This decision was made to show that the United States intended to preserve space as a "sanctuary." It was essential that space not become another testing ground for weapons so as to safeguard its use for observation, which Pentagon experts felt was crucial.[8] The decision signaling the beginning of the American civilian space program was made in July 1955. The first American artificial earth satellite, which was to be launched by the United States in 1958 on the occasion of the International Geophysical Year, would signal the beginning of a "peaceful space" era. But this first venture into space required the development of a launch vehicle. The sound approach was to make use of a technological development that had already been progressing spectacularly for several years: ballistic missiles. Intense competition soon developed to build the first launcher. The Army wanted to use a modified version of its Jupiter missile. The Air Force preferred its Thor missile, and gave the job of modifying it to Douglas Aircraft. The Office of Naval Research contracted Martin to develop the Vanguard, the first civilian launch vehicle, from the Navy's Viking missile.

It was a Jupiter-C rocket that, on January 31, 1958, placed the first American satellite, Explorer I, into orbit nearly four months after the Soviets' space premiere.[9] The event brought a sigh of relief in the US government after the spectacular failure of the third Vanguard launch on December 6, 1957, a month and a half after the successful launch of Sputnik 2.

The Soviet Union's progress led to acceleration of the American space program in both civilian and military sectors. The National Aeronautics and Space Administration was created on October 1, 1958, with a broad mandate to develop civilian space activities.[10] On October 11, using a Thor-Able rocket, the Pioneer I probe was launched toward the moon, which it "flew by" at 71,300 miles—a premiere that gave the United States the satisfaction, if only symbolic, of having regained the initiative in space.

On May 24, 1960, an Atlas-Agena rocket placed the first Air Force satellite, Midas 2, in orbit. It was designed to test an infrared device for detecting enemy rocket launchings, opening an area of research that 25 years later would still be important.

Progress in space research was spectacular on all fronts. By the mid-1960s, the Air Force and the Central Intelligence Agency possessed the technical capability to transmit satellite photographs of Soviet territory with a resolution of 3 meters. Progress in multi-spectrum photography made it possible to determine precisely the light-absorption frequencies of various objects, thus allowing the identification of military equipment hidden beneath various kinds of camouflage and even the monitoring of the productivity of Soviet harvests.[11]

Media coverage of the Apollo program focused on the many technological firsts brought about by civilian space exploration. The two roads now open for space development—civilian and military—were closely related as far as the technology was concerned. In fact, whith few exceptions, NASA's and the Pentagon's contractors were one and the same.

In relative decline during the 1970s, military space research was given renewed priority during the 1980s. SDI opened a new dimension for it. In 1986 the Pentagon's space budget was twice as large as NASA's.[12]

One of the last-born of the great American technological projects was announced by President Reagan in his State of the Union message in 1985: "NASP," the National Aerospace Plane.[13] The long-term objective was to produce a hypersonic aircraft capable of reaching Mach 25 (25 times the speed of sound)—the speed required to reach orbit. The "Orient Express," the commercial version, would carry passengers from New York to Melbourne in 2 hours by the beginning of the 21st century. The military version of NASP would be a transatmospheric interceptor flying between Mach 5 and Mach 10. But the main goal of the program would be to develop a successor to the space shuttle which would take off

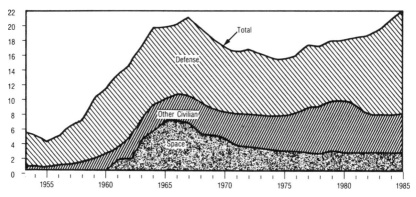

Figure 1 Evolution of the federal research and development budget (in millions of 1972 dollars). Source: NSF/OCDE.

from a runway like an ordinary airplane and then accelerate to orbital velocity with rockets. NASP was expected to provide access to space much less expensively than the shuttle. The multiple objectives of the program explain why its budget—$260 million a year since 1986—is shared by the Pentagon and NASA. A prototype, the X-30, capable of Mach 5, could fly by 1995. The objectives of NASP are controversial, and at present it remains a long-term research program for developing and testing the basic technologies needed for hypersonic flight: combined propulsion modes, advanced heat-resistant materials, and computerized studies of optimum profiles for very high velocities.[14]

A High Concentration of Federal R&D Resources

The continued development of military and space technologies has come about at the cost of unprecedented growth in federal expenditures for research and development. These expenditures peaked in 1967, the time of the highest spending on the Apollo program. In that year military and space R&D disbursements totaled $13.5 billion. They declined to no more than 60 percent of this amount in 1976 (see figure 1). Public investment in military and space research rose during the 1980s to a level comparable with that at the end of the 1960s. The 1985 budget exceeded that for 1967, in constant dollars, and again represented nearly 80 percent of the federal government's total R&D expenditures.[15]

Table 1
Distribution of federal research-and-development funds (percentages).

	Aerospace	Electronics and telecommunications	Other
1957	52.5	27.8	19.7
1969	53.5	28.3	18.2
1981	51.6	24.3	24.1
1988	55.1	23.7	21.2

On a sectoral level, these public expenditures are highly concentrated. Two large industrial sectors—the aerospace sector and the electrical, electronics, and telecommunications sector—have for 30 years been receiving an average of 80 percent of federal R&D funds, as table 1 shows.[16]

In the case of aerospace, 1988 federal contributions represent 79 percent of the R&D resources available for the entire industry, including civilian and military activities. This number illustrates the importance of government involvement in technological development in this sector. The manufacturers' share comes essentially from large aeronautics companies that are active in the civilian area, such as Boeing and McDonnell Douglas. Some armaments manufacturers also finance a small portion of their research themselves. However, a significant amount of this money comes indirectly from public funds, by virtue of a rule in Pentagon contracts that calls for reimbursing contractors for as much as 5 percent of the contract price for their own research outlays connected with it.[17]

The Key Role of Government Procurement

The federal government's involvement in the military and space fields is not at all limited to financing technological development. The particular weight these sectors carry in the economy comes from the industrial production of prototypes emerging from R&D. While the government's financial contribution is decisive in developing these new technologies, it is the purchase of a certain number of units of a given item of equipment that makes it an industrial reality. It would be pointless to undertake research if it were not to lead to products with some economic or military usefulness. But in contrast with most other industries, where hundreds or thousands of customers build the demand, thereby deciding which products will be developed, here the government-as-customer, via the Pentagon or NASA, alone determines the "usefulness" and hence the selection of military and space products. In these sectors it is up to the administrative departments alone to identify and call for the technologies they need for their respective missions, within the budgetary framework established by the government and by Congress.

This situation sets the armaments and space industries apart from most other high-tech sectors. It leads to a special type of company organization, and to supplier-customer relations to be found in no other industrial sector. It is a singularity worthy of special attention.

Pentagon expenditures vary greatly, depending on external crises or conflicts in which the United States may be involved. They have gone through three periods of major growth since World War II. The first was during the Korean war, the second during the Vietnam war, and the third has been underway since 1980. Depending on the period, the amount of public money injected into the US economy has represented between 5.5 percent and 13.8 percent of the gross national product. The latter percentage was reached in 1953 during the height of the Korean war. After falling to its lowest level toward the end of the 1970s, the portion of the GNP devoted to defense again reached 7 percent in 1985. A good one-third of these expenditures correspond to industrial equipment procurement. The rest is recycled through the economy in the form of salaries for the armed forces and various operating expenses. The armament industry employs nearly 3 million people, but in all more than 7 million jobs are directly or indirectly related to Department of Defense activities.[18]

The impact of these expenditures is particularly visible at a regional level, because of their geographic concentration. In 1986 firms in four states—California, Texas, New York, and Massachusetts—were sharing 42 percent of Pentagon contracts. This injection of public money figured strongly in determining the industrial characters of certain geographic areas, particularly in the southern and western states.[19] The Los Angeles area is a good example. The aircraft plants established there in the 1930s saw considerable development during World War II and in the 1950s. During that period, population growth in the area was directly related to growth in the volume of Pentagon contracts. New manufacturers of military aircraft came into the area during the 1960s and the 1970s. Most of the large companies involved in this sector today have part of their activity near Los Angeles. They include Lockheed, Hughes, Rockwell, Northrop, McDonnell Douglas, TRW, and General Dynamics, as well as a host of subcontractors.[20] In the 1960s, although firms working directly in defense represented only 8 percent of California's employment, when indirect employment was considered defense was estimated to involve 40 percent of the state's working population.[21]

In the space sector, public contracts have less impact. There is a difference of an order of magnitude in volume between Pentagon contracts and NASA contracts to industry. The chief business of defense contractors is manufacturing new equipment for the Air Force or the Navy, often in large series. In the space field, on the other hand, the technologies developed to meet NASA's requirements generally lead only to the production of single items or small batches. This explains why, even at the time of the space program's greatest development, in the late 1960s, its global impact on the economy was much less significant than that of military expenditures. In 1967 the latter amounted to $70 billion, including more than one-third in direct procurement of manufactured goods, against only $5 billion for the space sector.

The Pentagon's industrial purchases are based on the government's priorities in regard to modernizing or renewing weapon systems. In 1984 these purchases, with a total volume of $86.3 billion, broke down as follows[22]:

• aircraft and aeronautical equipment: $29.2 billion
• missiles and space technology: $19.7 billion
• electronics and communications equipment: $16.7 billion
• ships: $9.4 billion
• land vehicles: $5 billion
• munitions and miscellaneous weapons: $6.4 billion.

These figures underscore the considerable weight of the aeronautics sector in defense procurement over the last 20 years. Also apparent is the importance of electronic equipment, an increasing area of military spending. Electronic components are widely used by all manufacturers of weapons equipment and systems, which increases their weight in defense expenditures. In 1984 total Defense Department spending in this sector was roughly estimated at $30 billion.[23]

The Pentagon's priorities are also reflected in the list of firms holding its contracts. In 1985 the Pentagon's chief suppliers were, in decreasing order, McDonnell Douglas ($8.9 billion in public orders), General Dynamics ($7.4 billion), Rockwell International ($6.3 billion), General Electric ($5.9 billion), Boeing ($5.4 billion), Lockheed ($5.1 billion), and Hughes ($3.2 billion).

The Pentagon's contracts are shared among a thousand or so major contractors, and some 150,000 subcontractors (including a large number of small and medium-size firms). Nearly 15 million contracts or miscellaneous purchase orders are signed annually—roughly 50,000 a day. To handle this gigantic management task, along with its defense-related missions, the Department of Defense employs approximately 2 million persons. The complexity and massiveness of this organization, whose priorities are not always clearly defined and whose orientations are continually changing, caused MIT economics professor Lester Thurow to state that the Pentagon is the largest planned economy in the world outside of the Soviet Union.[24]

The planning is truly difficult, too. Equipment choice is dictated principally by the criterion of military effectiveness. This depends on performance, and improving performance generally means increasing complexity, which in turn means higher costs. Not surprisingly, the cost of a radar unit for a combat plane increased tenfold between 1962 and 1971.[25] But it is also clear that this device's performance was greatly improved between those two dates. We can therefore understand the chronic problem facing the Defense Department's procurement officials. In the context of growing pressure for ever more complex and expensive equipment, tradeoffs must constantly be made between adequate performance and the need to stay within the available budget.

A Special Relationship between the Suppliers and the Customer

The result of the constant tradeoffs between price and performance is a very special relationship between the industrial suppliers and their customer, the government. During contract negotiations the latter will usually give priority to the satisfaction of narrowly defined technical specifications. The knowledge that the contractor has the technical competence required to meet these specifications will generally outweigh any concern for finding one able to do so more cheaply. Such a quest would often be in vain in any case, so specialized are the companies involved in defense contracting. This results in a highly fragmented market pattern, with a small number of companies in each segment, and technical barriers to entry by newcomers. In 1980 the four most important firms had a 97 percent market share for solid-fuel boosters, 96 percent for inertial navigation systems, 93 percent for jet aircraft engines, 99 percent for nuclear submarines, and 74 percent for combat vehicles.[26] And for some particular items they are pure monopolies—for example, General Dynamics is the sole supplier of Trident strategic submarines.

Cutthroat battles may still arise among two or three large companies bidding competitively. Winning a contract for hundreds of millions or even billions of dollars may be vital for a giant firm in the defense sector. Once the contract has been awarded, however, the Pentagon often has no choice but to accept the supplier's demands. This is particularly true if the contract was made on a "cost-plus" basis—current practice for a large proportion of new equipment orders. In that case, the supplier is not held responsible for cost overruns.[27] Under these conditions there is no risk for the supplier, since the government picks up the tab for all expenses.

Relations between the Pentagon and its suppliers are facilitated by the fact that individuals move easily between government and industry, filling key jobs in both at different points in their career. A good example is provided by General Dynamics. Its president, James Beggs, was NASA Administrator from 1981 to 1985. George Sawyer, a former Assistant Secretary of Defense, joined the company as vice-president in 1983. This is not unusual. Rockwell hired

Richard De Lauer, Deputy Secretary of Defense for Procurement during Ronald Reagan's first term.[28]

Together these characteristics ensure great stability for the firms centered around military and space programs. In a comparison between the top twelve Pentagon suppliers in 1970 and 1985, there are only three new names—two of which were already on the list in 1970, but lower.[29]

The Pentagon's system for awarding contracts has been the subject of much criticism. The major reproach is that it produces considerable inefficiency. Low productivity and mediocre quality control do in fact characterize many of the companies that work for the Department of Defense. "For years," notes the *Wall Street Journal*, "profits for Pentagon suppliers were based on costs. As a result, these companies had little incentive to modernize their production facilities, since any savings could mean a renegotiated contract for a lower price, hence lower profits." Greater openness to competition for a larger proportion of defense contracts in recent years seems to have introduced an additional uncertainty factor, leading many manufacturers to postpone decisions to modernize production facilities.[30] Part of the problem stems from the yearly nature of military budgets, which precludes any medium-term production planning for suppliers. The result is that many firms in the defense sector produce highly sophisticated equipment using production techniques that may be 50 years old.[31]

The scandals that periodically erupt in the press about certain Pentagon contracts (Lockheed during the 1970s, General Dynamics and Rockwell during the 1980s), with accusations of padded expenses or misappropriated public money, are other signs of weakness in a system that no one has seriously attempted to revise. According to a 1985 poll, 70 percent of respondents felt that padding and cost overruns were frequent practices among the Defense Department's suppliers, and 62 percent felt that any company caught defrauding the government should be banned from the market.[32]

In practice it is not that simple. As was emphasized by the newspaper that conducted the above-mentioned poll during a scandal over General Dynamics' having billed cost overruns to the Department of Defense, over the years that company had made itself indispensable to the defense of the United States: "Its F-16 is one of the combat aircraft most esteemed by Pentagon and NATO buyers. Its Tomahawk cruise missiles and its M-1 Abrams tank have become essential in the arms race with the Soviet Union . . . and it is the only company capable of building the Trident nuclear submarine."[33]

In an area where comparative measurements of performance are difficult, or even impossible (as in the case of weapon systems intended to be used only once), it is hard to make a global judgment about the efficiency of the system that produces technological goods for the military and the space program in the United States or in many other countries. Beyond what can only be a value judgment, this system has unquestionable merits: it has consistently served to advance the frontiers of technical knowledge, it has produced some of the most advanced technologies ever developed, and it is a powerful influence on technological innovation in America.

Two Cultures

4

When Ruben Mettler took over as the head of TRW, in 1977, his ambition was to make the company into a leader in the electronics industry. Ten years later this goal seemed to have been met. In 1986, 56 percent of the company's $3.4 billion in sales came from this sector. Nevertheless, there was a shadow in the picture. Although TRW had managed brilliantly to diversify into military electronics, it had failed to obtain a single civilian contract.[1] A few years ago, after hesitating at length, it decided to get out of the market for 64K RAM computer memories, and then from the telecommunications satellite market. In 1979 the company had signed an agreement with the Japanese firm Fujitsu to develop activities in the field of digital switching, a rapidly growing branch of telecommunications. The adventure had ended two years later with a loss of $35 million and with the company's withdrawal from this activity.[2]

TRW is not the only large defense contractor to have had trouble entering civilian markets. Rockwell, one of the principal contractors on the Apollo program, decided in 1973 to enter consumer electronics. For this purpose it purchased Admiral. Under Rockwell's leadership, Admiral soon introduced what was then considered to be the "Cadillac of TV sets," incorporating all the latest improvements in video technology. Unfortunately, the product was much too expensive. Rockwell admitted its mistake and sold Admiral in 1979.[3]

In the early 1980s, McDonnell Douglas decided to make a major diversification effort in the field of information technologies and systems. The company undertook to create a time-sharing computer system and automated manufacturing units. In 1984 its president announced a goal of $4 billion in revenues in this area by 1990, and 7 percent profits after taxes. But an entirely different result soon emerged. The McDonnell Douglas Information Systems affiliate lost $109 million in 1985, twice as much as in the preceding year, for a turnover of $1.1 billion.[4] A more recent attempt to apply McDonnell Douglas' experience in the space sector to manufacturing commercial products could also end in failure. In 1980 McDonnell entered into an association with Johnson & Johnson with the goal of producting a hormone under weightless conditions in space. An experimental electrophoresis production unit was developed and tested on several shuttle flights. However, as McDonnell was sending its machinery into space, effectively proving that it was possible to manufacture the hormone erythropoietine under gravity-free conditions, a small California biotechnology firm called Amgen succeeded in producing it on earth through genetic engineering. "Between what McDonnell did and what we did," Amgen's manager said recently, "there is as much difference as between producing a book by hand during the Middle Ages and reproducing it on a copy machine." Johnson & Johnson dissolved the partnership with McDonnell to join up with the biotechnology firm.[5]

Will drugs one day be manufactured in space? Possibly, but McDonnell's misadventure shows how difficult it is to apply the sophisticated technologies of the space program to rapidly changing commercial markets. This characteristic appears throughout the military sector, where products using the most advanced technologies rarely find direct applications in the civilian market.

Although it is possible to name some notable exceptions to this rule, there are many significant factors that contribute to the distinction between activities related to large-scale technical programs nurtured by public funds and those aimed at developing commercially viable technologies. In fact, these sectors are not simply separated by a barrier that makes access between them difficults; there are, rather, two universes, foreign to each other, operating on different bases, with different objectives and motivations.

Two Approaches to Innovation

One difference concerns the manner of assuming the risks tied to the development of a new product. For a company in the commercial sector, introducing a new product always means an adventure with high financial risk. Most often the company will assume the financial burden itself. This is why, before deciding to finance the development of a product, a company must be sure that it corresponds to a solvent demand, and that there is a reasonably good chance to finalize its development within a limited time. Then, if the developed product lives up to expectations, there still remains the task of positioning it in the market. How many units should be produced, and at what price? The answer to this question will depend on a number of parameters, among which are production costs and commercial strategy. Will it be a top-of-the-line product with unique characteristics, or a mass-market product? If the latter, the company's principal concern will be to minimize production costs in order to increase sales volume. For a company in the commercial sector, the size and efficiency of its productive investments are essential factors in the decision, and will seal the new product's fate.

In contrast, a company working for the Department of Defense runs only limited risks. More often than not, if unforeseen problems arise, the cost of developing the new product will be covered by the customer. Although the contractor's production costs may not be a negligible factor, they will generally not be a determining element in his contractual relationship with his customer. The most important issue in their transaction has to do with the technical specifications to which the contractor is committed, along with the future product's performance. In most cases it will be produced in a limited series, if not as a single unit. Manufacturing conditions are generally a secondary consideration. Moreover, a company working for the Department of Defense will generally not know until some later stage the volume of equipment being ordered, since this depends on a complex budgeting process involving various partners, levels of decision, and other multiple influences not taken into account by the usual type of economic calculations.[6]

Whereas a company in the commercial sector must necessarily take account of the characteristics of similar products being manufactured in Europe or Asia, one in the military sector is usually insulated from international competition. In fact, the Department of Defense is required by law to buy American, on the condition that the domestic supplier's price may not be more than 50 percent higher than the foreign supplier's. As Lester Thurow has observed: "Military procurement is a captive market for American industry. The Pentagon does not invite foreign companies to bid on military equipment. If it did, all of the Navy's ships would be built in Japan or Korea."[7]

A firm in the commercial sector must, for the sake of survival, be constantly watching market signs; a defense contractor needs to be concerned with extending his influence within the bureaucratic structure of the Pentagon, in order to be an early bidder on new projects. This requirement can have a profound effect on a company's organization. As has been noted by Jay Stowsky, a researcher at the University of California at Berkeley, "The contracting firm begins to mimic the bureaucratic structure of the Pentagon itself in order to facilitate effective communication, isolating itself, in the process, from channels of communication with civilian customers."[8] The activities of the two firms lead to differences in organizational logic, to different corporate philosophies, and to correspondingly distinct concepts of innovation.

Innovation is obligatory for all companies dealing with competitive markets. To different degrees, all must attend to the creation of new products or services to meet their customers' changing needs and expectations. They are also forced to innovate in order to improve productivity. This requirement is all the more compelling as a firm's environment becomes increasingly competitive. The stronger the competition, the more compelling the need to innovate with respect to manufacturing processes.

The means of acquiring technology and know-how may vary, depending on the sector and the size of the firm. The smaller its size, the more dependent it will be on outside acquisitions of patents and know-how. The larger the company, the more endogenous resources it is likely to be able to devote to R&D. Such an effort is all the more necessary if the firm belongs to a rapidly changing sector. This need increases concomitantly with expansion of the scientific and technological base required for industrial development. In 1987 American industry devoted nearly $60 billion of its own resources to R&D—four times as much as in 1960 in constant dollars.[9]

The relationship between a company's investment in research and its ability to bring new products or processes to the marketplace is far from immediate, however. The process of innovation is rarely linear. Few successful products result directly from an innovation process moving step by step from basic research, to applied research, to product development, to production and marketing. Numerous investigations of American firms in different industrial sectors have shown that, on the contrary, most firms innovate—decide to create new products or manufacturing processes—on the basis of demand analysis or external contacts.[10]

The characteristics of the market in which a company is operating also have a profound impact on its innovation process. In new industrial sectors, *product* innovations generally dominate. There is greater emphasis on *processes* once the market has reached a certain degree of maturity.[11] The circumstances leading to innovation are different in every case, strengthening relations with the marketing department in the first instance and with production shops in the second. In practice the process of innovation in a commercial sector results from complex and multiple interactions among three distinct worlds: research laboratories, production units, and the market.[12]

A very different innovation process characterizes companies in the military market. It stems from the particular concerns of the customer—in other words, the Department of Defense.

First of all, more than anything else, this particular customer wants the most advanced technology, in order to maintain the United States' position in the strategic arms race.[13] This concern explains the huge sums spent on R&D in comparison with the overall industrial revenues in the armaments sector. The ratio is 25 percent, versus only 10 percent in the computer industry and 12 percent in semiconductors. This race for equipment sophistication and performance leads to heightened product innovation. Companies are organized to build prototypes or experimental weapon systems, for which the development and the testing absorb a large proportion of the available R&D funds. Narrow specialization is often required for these tasks. For each category of equipment, the Pentagon usually selects a few firms whose technical competence has been proved—this explains the relative concentration of R&D

funds in the defense industry, more than half of which are shared by ten companies. This oligopolistic industrial structure, in turn, encourages a race for equipment sophistication. Widening their technical lead is the chief means available to these firms for maintaining their influence with the Pentagon and preventing competitors from entering their field.[14]

The second concern of the Department of Defense is to prevent the results of the research it is underwriting from falling into enemy hands. This characteristic of military research profoundly affects the mechanisms of innovation. In contrast with the commercial sector, where technical progress is in large part the result of external interactions, technological development is mostly endogenous for a Pentagon contractor.

In fact, firms in these two sectors are characterized by two very different processes of innovation. The difference between these two worlds is not only technical and organizational but cultural. We are dealing with two different technical cultures.

The first culture generates products whose filter is the market place. The condition for survival for companies in this arena is their ability to produce standardized products for a dispersed customer base that are less expensive and/or of better quality than their competitors'. Here, product price is an important criterion for the customer's choice of supplier, and often the most determining one.

The second culture is one in which custom-made and highly technical products are made individually, or in small series, for a single customer—usually the government—whose principal concern is that the products meet predetermined levels of performance. The rule of the game in this world is to anticipate and satisfy this specific customer's preferences. The financial risk run by a company involved in this field is generally minimal. Although there may still be some competition, company specialization and long-standing working ties between equipment suppliers and buyers limit its effects.

The first culture corresponds to the modern transposition of a process that has characterized change and adaptation in industry since the beginning of the first industrial revolution. From Robert Fulton to Henry Ford, from Thomas Edison to Steve Jobs, technical innovation has been for American businesses a source of constant opportunities and renewal. This first culture therefore has very old roots. It illustrates the adaptation to the innate constraints of the US high-technology sector, of the classic model of industrial production in a free-market economy. The second culture appeared much more recently. It is the result of the dominant influence gained by the government over technological production during and after World War II—an influence that, in various forms, has since become perennial, as we have seen.

In many ways these two cultures are opposites. At the same time, they both make an essential contribution to the development of American high technology. This apposition does not mean they are alien to one another. A debate over their relationship—their synergies and their antagonisms—has continued through the whole history of American technological development, and is more relevant today than ever before. Can the second culture, whose goal is to provide the technologies needed for national security and other national goals, have any impact on the first one, whose aim is technological development for commercial purposes? This question is central to an understanding of the government's role in technical innovation and industrial development in the United States. Although it may oversimplify a somewhat more complex situation, making a distinction between these two cultures provides a tool for evaluating that role, and for understanding the underlying springs and mechanisms of technological development in the United States.

Common Roots

The split and the apparent opposition between these two technological cultures did not always exist in America. On the contrary, during the period following World War II there was a close symbiosis between them. A quick analysis of three branches of industry—semiconductors, data processing, and aeronautics—will help us understand how and why.

No sooner had William Shockley invented the transistor than the Department of Defense became interested in it. If fragile, unreliable vacuum tubes could be replaced with small, sturdy, low-energy-consumption semiconductors, it would be a tremendous advantage in military electronics. In 1952 the Air Force estimated that 40 percent of all its equipment could be transistorized, producing reductions of 20 percent in size and 25 percent in weight and a reduction in failures of 40 percent.[15] This explains why the Pentagon financed research in semiconductors on such a massive scale. The funds went mostly to the traditional suppliers of tubes: RCA, Sylvania, Raytheon, Philco, and General Electric. A few newcomers to the industry nevertheless managed to obtain some government funds. One was Texas Instruments, which in 1952 made an important strategic choice in hiring Gordon Teal, a Bell Labs alumnus. The early transistors had been made of germanium, a very expensive material. Teal was the first to consider silicon as an alternative. This material soon became a top research priority for the Pentagon—not because the military had foreseen the spectacular way in which silicon would later come to dominate the industry, but simply because silicon made it possible to use much higher operating temperatures than with germanium. The hiring of Gordon Teal not only helped Texas Instruments to win Defense Department research contracts; it also gave the company an important advantage over its competitors in the contest to become the Pentagon's prime supplier. Texas Instruments foresaw the large role military procurement would play in the fledgling semiconductor industry until such time as the civilian market took off.[16]

Procurement orders in 1955 represented nearly 40 percent of all purchases from the semiconductor industry. This proportion was even higher for another innovation, one that would mark the first "technological leap" in the profession after the invention of the transistor: the development of the first integrated circuits by Texas Instruments and Fairchild in 1959. The principle of the integrated circuit was conceived by Jack Kilby, who was working for Texas Instruments.[17] He had the idea of making other components for electronic circuits, such as resistors and condensers, out of the same material as transistors, so that they could be integrated on the same silicon chip. The space saved was considerable. Integrated circuits were soon being used in missile guidance systems, and progressively in all military applications where miniaturization played a decisive role.

But civilian needs for integrated circuits were slower to arise, because the circuits were still expensive and because they implied major organizational changes in the industries that would use them.[18] In 1963, when the circuits still cost $50 apiece, practically all of the industry's turnover, nearly $4 million, came from government procurement. Two years later, when the selling price had fallen to $9 per circuit, the government share of business represented no more than 75 percent of a market now worth $80 million. The Defense Department provided substantial resources for developing the technology. By 1959 Texas Instruments had been awarded $1.15 million in military research funds, and the following year the firm won a contract for $2.1 million to study the mass production of integrated circuits.[19] Nevertheless, as in the case of the transistor, public money did not account for the basic discovery of integrated circuits. Above and beyond its contribution to the R&D effort, the Defense Department stimulated industrial development through large orders, making it possible to expand production at a time when civilian demand was still extremely weak.

A similar pattern characterized the first stages in the development of the computer industry.

Science historians agree that the common ancestor of modern computers was ENIAC (Electronic Numerical Integrator and Calculator), developed in 1945 at the University of Pennsylvania. Unlike its predecessors, ENIAC was fully electronic, and its calculating speed was several hundred times faster than the best of the electromechanical calculators. In addition, this new machine introduced an important innovation: logical instructions stored in memory could be modified without using thousands of switches.[20] The Defense Department and several other government agencies had financed ENIAC, for its aims were too close to basic research to interest industry. Nevertheless, six months later representatives from several dozen companies, including some of the future computer manufacturers and a large number of defense contractors, participated in a four-day seminar at Harvard University during which a presentation of ENIAC's achievements was made. The audience showed great interest. This marked the effective beginning of the development of a new industry.

Through the mid-1950s, the demand for electronic calculators was limited to government laboratories working on nuclear weapons or missiles and to other public agencies, such as the Weather Bureau and the Coast and Geodetic Survey.[21] This explains why most of the computers of the time—the so-called first generation—

were developed by the industry with the aid of government sub-sidies. This was notably the case for the Burroughs E-103, General Electric's ERMA, RCA's BIZMAC, and Remington Rand's Univac 1 and 2. IBM also won a defense contract to develop STRETCH— the first version of the IBM 7030—a machine 100 times more powerful than the IBM 704, which had been ordered by the Los Alamos laboratory in 1956.[22] Experience acquired on the STRETCH project enabled IBM to develop its 7090 series, which in 1960 marked the company's entry into the market for fully transistorized, second-generation computers. The industry changed markedly with the arrival of these new products. Many new com-panies were created as civilian orders for computers began. Control Data Corporation and Digital Equipment, two firms that would become powers in the field, were created in 1957. Control Data received its first Pentagon contracts to develop its 1604 model. The development costs for the more powerful 3600, whose introduction was announced for 1962, were entirely covered by the Air Force. The first machine went to Livermore Laboratory in 1963.

The history of the Digital Equipment Corporation is insepa-rable from that of Route 128, the beltway outside Boston along which new high-technology companies began to gather in the 1950s. Digital Equipment was their prototype. Kenneth Olson and Harlan Anderson, DEC's founders, came from MIT's Lincoln Lab-oratory, where they had worked on SAGE, one of the first com-puter projects. SAGE, financed by the Air Force, had as its objectives the interpretation of radar data and the organization of retaliation in case of aerial attack. Olson and Anderson started their company in 1957 with $70,000 provided by General Georges Doriot, a Frenchman who created the American Research and Development Corporation, one of the first venture-capital firms in the Boston area.[23] Unlike many computer firms in the 1950s, Digital Equipment had little public funding for the development of its products. It was the first company to consider developing com-puters for private customers. From the outset the company devel-oped an original product line: the PDP series, forerunners of the minicomputers. It was also, in the 1960s, the first manufacturer to produce an interactive computer that allowed the user to enter data via a terminal with a screen and a keyboard. Much later, in the 1980s, Digital Equipment was the first to introduce high-speed computer networks, using its VAX system. In 1982 its gross income reached $4 billion.[24]

Although from the beginning it had adopted a strategy aimed at commercial markets, Digital Equipment did benefit, like other computer manufacturers during the late 1950s and the early 1960s, from government procurement. But private computer markets took off very soon, filling in where government contracts had left off. By 1962 government procurement represented only 48 percent of the demand for computers, versus 72 percent three years earlier.[25]

A similar phenomenon occurred during the 1960s in certain high-technology markets. After startups made possible by defense contracts, the industry's development was determined by expanding civilian demand. The military demand for semiconductors declined from 50 percent of the market in 1960 to 25 percent in 1972, and to 10 percent in 1979.[26]

This evolution had a profound impact on the structure of the industry. During the 1950s most high-tech companies were organized along the lines of the second-culture model, their prime objective being to "push" technological development so as to take advantage of military funds. This was clearly the dominant innovative pattern in the computer industry during the 1950s.[27] At the beginning of the 1960s, new firms were being created with an entirely different goal: to penetrate commercial markets. As a result, the computer market was for several years in a hybrid situation. On the one hand, there was a fast-changing technology, in many ways still experimental, with little product standardization, which tended to generate a second-culture developmental mode. On the other hand, increasing competition was causing companies to adopt structures based on a logic that was more and more that of the first culture. Longer-established firms were faced with the strategic choice of preserving their share of government procurement business or seeking to take advantage of the newly expanding civilian markets. For many of them this meant either reinforcing a position in the second culture or swinging to the logic of the first.

This choice became all the more inevitable as products for military markets evolved toward greater complexity than civilian products. In addition, the secrecy surrounding technological development in the second culture quickly became a burden for firms that had moved toward the first culture. Beginning in the 1960s

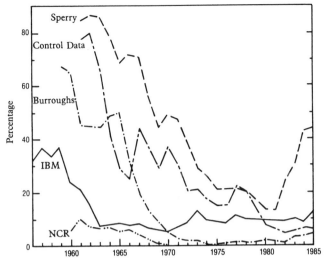

Figure 2   Share of R&D resources provided by Department of Defense to various data-processing companies. Source: *Targeting the Computer* (Brookings Institution, 1987).

this issue played an important role in the strategic decisions being made by certain firms, such as IBM. While taking advantage of expanding civilian markets, IBM decided in 1961 to reduce its involvement in military R&D contracts. A similar step was taken a few years later by Burroughs, Control Data, and Sperry[28] (figure 2).

These decisions had important consequences for the financing of industrial research. In order to keep up with ever-quickening face of technical change, companies in the first culture had no choice but to increase the share of their own resources devoted to research and technological development. While the volume of public funds invested in industrial R&D in 1981 was comparable to its 1960 level (in constant dollars), the amounts the industry devoted to this activity tripled during that period. In 1960 nearly 60 percent of companies' resources for R&D came from the federal budget. In 1981 it had fallen to 31 percent. By this time the Pentagon's and NASA's large projects were providing only a minor part of the resources being devoted to technological development in American industry.

By the end of the 1960s, firms belonging to the first culture had for the most part considerably pruned or cut their roots in the second. This was the period in which the first culture, through a flood of new companies, was beginnning to express its basic characteristics along Route 128 and in Silicon Valley. As they turned toward different objectives and markets, propelled by different innovative rationales, with the greater part of their technological development covered by resources of different origin, the firms in the two cultures were becoming increasingly divergent in their concerns and their activities.

A rift was growing between the cultures. It would widen progressively in most of the high-technology sectors except aeronautics, an important exception to the evolution I have just described.

When the Second Culture Looks to the Marketplace

As it did in the other leading sectors I have reviewed, the government played a leading role in bolstering commercial aeronautics; however, in contrast to the cases of semiconductors and data processing, its influence here continued long after the industry had reached maturity. A brief historical review will help us understand the reasons for this singularity.

The origins of civilian aeronautics go back to the period before World War II. The circumstances of its development are unique in the annals of American industry because of the key role played by a government agency: NACA (the National Advisory Committee on Aeronautics), the forerunner of NASA. Between 1920 and 1935 NACA financed the bulk of research in airframes and engines and made the first wind tunnels and testing facilities available to the industry. Beginning in 1935 it increasingly served the needs of the military aeronautics sector, although it continued to exert great influence over the technologies required by the civilian sector.[29] In fact, from then on, military needs would lie at the root of most innovations in aeronautics, even though much of the technology developed to meet them would later be transposed to the commercial sector. This affiliation is particularly striking in the case of engines, where it still persists today. The list of innovations developed for military prototypes and later applied to civilian aircraft includes the air-cooled engine for the DC-3 (the best-selling civilian airliner through the early 1950s) and the high-bypass turbofan (which powers the large modern jets, including the Boeing 747, the Lockheed L-1011, and the DC-10).[30] The jet engine, a major innovation that made commercial jets possible in the late 1950s, was developed in the United States by General Electric and Pratt & Whitney from a British patent used to fulfill US Air Force requirements during and shortly after the war.

Not until after World War II did significant connections appear between military and civilian airframe technologies. With the introduction of jet aircraft, civilian manufacturers took advantage of progress made in the conception and design of military airframes. In some cases the similarity of the airframes developed to meet military and commercial requirements led to considerable savings in the development costs of civilian aircraft. The best-known example is the Boeing 707 (the first American commercial jet airliner, introduced shortly after the British Comet and the French Caravelle), whose airframe was directly derived from the KC-135 tanker, built by Boeing for the Air Force. The 707's design was so close to the KC-135's that the first prototypes of the airliner were devoid of windows, like the tanker. Boeing's savings on this occasion were certainly substantial; a difference of more than $130 million appears between the losses that Boeing wrote off for the development cost of its 707 and those that its competitor McDonnell Douglas—which did not have a similar advantage— wrote off for the development cost of its DC-8.[31]

There are other examples of how Air Force procurement has enabled manufacturers to develop airframes or components and their associated manufacturing tools, and then use the know-how in the civilian domain.[32] Such transfers became increasingly important as the cost of developing civilian aircraft grew, increasing at approximately the same rate as the cost for military aircraft. In the 1930s, it had cost $300,000 to develop the DC-3. Immediately after the war, the cost of developing a new aircraft had risen to tens of millions of dollars ($14 million for the DC-6 and $29 million for the two B-47 prototypes). By the end of the 1950s it exceeded $100 million ($112 million for the DC-8 and $468 million for the McDonnell Phantom). It had reached $1 billion by the late 1960s (DC-10, Boeing 747).

These two factors—large investments that can be recovered only over a very long period and manufacturers' need for a strong technological base tied to the military sector—explain much about the structure of the aeronautics industry, in which a small number of firms derive their revenues from both civilian and military activities.[33]

Of the four companies competing in the civilian aircraft market in the 1940s and 1950s—Douglas, Lockheed, Convair, and Martin—only Douglas remained after Boeing's spectacular comeback in civil aeronautics in 1958. And Douglas, a giant of commercial aviation before World War II, was able to survive only thanks to its 1967 merger with the military aircraft manufacturer McDonnell.[34] In 1958, the year the 707 was introduced, Boeing's military activities represented 96 percent of its revenues. This company was one of the major suppliers for the US Air Force during the postwar period. It produced the B-29, B-47, and B-52 strategic bombers.[35] The share of its defense-related activities began to decrease in the late 1960s, however, because of the commercial success of the 747 (introduced in 1969), the 727, and the 737. By 1976 military procurement represented only 25 percent of Boeing's sales, versus 36 percent in 1969. However, this proportion again increased in the early 1980s, reaching 50 percent in 1985.[36]

As civilian markets expanded and diversified, manufacturers were able to make great gains in productivity. Modern assembly-line methods applied to aircraft manufacturing led to substantial economies of scale.[37] But, as Sussex University researcher Mary Kaldor has noted, the production mode and the logic of innovation in the aeronautics industry remained those of the second culture: ". . . the dominance of military thinking infected the approach to civilian design and marketing. The massive engineering teams which proliferate ideas, problems of technical coordination, undetected mistakes, cost and complexity, and the constant striving for periodic improvements in performance were all applied to commercial aircraft. For example, over a thousand people designed the Boeing 727."[38]

Aside from their technical affiliation, close similarities still exist today in the development process for civilian and military aircraft. The structure of demand—that is, the airline industry—contributed to perpetuating in the civilian sector the same model for innovation found in the military sphere. Up through 1978, when the deregulation of the airlines went into effect, purchases of civilian aircraft were in the hands of four airline companies. These firms played a crucial role in the innovative processes in civilian aeronautics. They controlled the competition, and they increased the stakes for manufacturers introducing new models. The competition could become extremely harsh, since the loss of market share represented by a large airline such as Pan Am might threaten the manufacturer's very existence.

A very special relationship, not unlike the one between the Pentagon and its suppliers, has arisen between aircraft builders and large airlines. The airlines play a decisive role in defining the characteristics of future aircraft. Like the buyers of defense equipment, they are focused on the equipment's technical characteristics and on its ability to meet certain performance requirements. These discussions are all the more important for the airline companies in that they are prisoners—at least they were through 1978—of the Federal Aviation Administration's tariff structure. Unable to compete on prices, the companies set themselves apart via service. Under such conditions, a new-model airplane, with all the latest improvements, is an important marketing argument.[39] This aspect of the situation explains why airlines still want aircraft that meet their specifications and needs as closely as possible. Although it uses modern mass-production techniques widely, the civilian aeronautics industry has remained a realm of semi-custom manufacturing.

This close liaison between aircraft manufacturers and airline companies requires both to maintain large teams of technicians and specialists. The civilian aeronautics sector, like the Pentagon's client industries, carries an extremely high negotiation and transaction cost.[40]

If the manufacturer is open to negotiating the technical characteristics of future aircraft with its airline customers, it also wants their involvement in sharing the financial risks of the operation. The manufacturer will therefore not undertake such costly development unless he has guaranteed sales for a certain number of the new aircraft. He may in fact abandon a project for which he feels he has inadequate purchaser guarantees. This is what happened when McDonnell Douglas considered building the DC-11. Delta Airlines actively encouraged the manufacturer to do so because it wanted an alternative to the Boeing 757. But McDonnell felt that a single company's commitment was not enough, and so chose not to build the new plane,[41] leaving Boeing with an almost total monopoly for its 757 and 767 models.

In spite of their desire to associate their customers as early as possible with the development of new airplanes, the manufacturers must nevertheless assume most of the financial risks. These development costs can be recovered only over a very long period, and only if several hundred aircraft are built. Consequently, introducing a new model of a commercial aircraft is a high-risk venture. This is what distinguishes civilian aircraft manufacturers from their defense counterparts.

For several reasons, however, the difference between them is less fundamental that it might appear at first glance.

First of all, what may be true with respect to a particular product (such as the DC-10, for which McDonnell Douglas received no development subsidies) may not be so for the industry as a whole. For 25 years the Department of Defense has consistently provided three-quarters of the research funds available to the aeronautics industry. Since most of this financing is exclusively for the development of military prototypes and equipment, manufacturers have a strong interest in tying the two together. For the industry as a whole, including both civilian and military aircraft manufacturing, the Department of Defense is therefore assuming most of the risks associated with technological development.

But, even more significant, it is the high rate of profitability shown by these companies for their military activity that limits the consequences of the risks they may take in the civilian sector. From 1965 to 1985, McDonnell Douglas showned a profit in the civilian sector only three times. This is what led David Mowery and Nathan Rosenberg, in their study of the government's role in civil aeronautics development,[42] to say that military procurement has consistently provided a regular source of profits enabling the large aircraft manufacturers to finance gambles in the commercial sector. Moreover, no matter how great the risk, such gambles have not really endangered the firms that have made them.

"The government simply cannot allow one of its major defense contractors to go bankrupt regardless of its commercial mistakes," says S. L. Carrol, an aeronautics-industry analyst.[43] The case of

Lockheed in 1971 proves this point. Lockheed, trying to make a comeback in civilian aviation with its new L-1011, was in deadly competition with the McDonnell Douglas DC-10. At stake was control of second place in the jumbo-jet market, behind Boeing. This violent competition, coupled with the collapse of Rolls-Royce, Lockheed's partner for the L-1011's engines, put Lockheed in a precarious financial situation. Lockheed's bankruptcy was prevented by a $250 million loan guaranteed by the federal government.[44] (The company has since abandoned the commercial aircraft market to concentrate on protected military business.)

These are far from the operating rules for most companies in the first culture. A commercial risk that cannot lead to bankruptcy in case of failure, a product-development process that can take advantage of technological transfers from government-sponsored R&D, a bilateral oligopoly between seller and buyer, and direct producer-customer dialogue all tend to narrow the gap between the operating rules and innovative processes of civilian aeronautics and those in the military field. They explain why aircraft manufacturers have a strong technical and financial interest in concurrently pursuing both types of activity.

Unlike the high-techology firms competing in commercial markets that I discussed earlier, civilian aeronautics manufacturers are involved in an activity that is (or at least was, through the late 1970s) firmly anchored in the second culture. The commercial dimension of the aeronautics industry is, in fact, less important than its other characteristics in defining its position with respect to the two-culture model.

The Sheltered Culture versus the Exposed Culture

This conclusion leads to an important remark about the differentiation we have thus far made between the two cultures. Remember that I am trying to delineate a boundary between two technical cultures, between two different processes of technological innovation within American industry. Reviewing the history of the aeronautics industry prior to airline deregulation, we see that the borderline is not to be found between firms operating in the government market and those operating in private ones, as I provisionally expressed it, but between different types of commercial activity. We can identify the firms that belong to the second culture by the nature of their products and by the logic of the competition in which they are involved. The majority of them make complex custom or semi-custom products for a limited number of large customers. Their activity requires heavy investments that can be recovered only over long periods. The logic of these firms' behavior is that of monopolistic competition, as economists understand it. Whereas in the first culture pricing is crucial to a company's efficiency, here competition is based on providing the unique technical features required by the customer. A producer will be competitively successful to the extent that he has built an effective communications interface with his customer, and to the extent that his expertise gives him a technical edge—not a price edge—over his competitors.

The example of the aircraft industry also suggests that companies selling such products may not be able to do so in the absence of externalities resulting from government intervention. Such externalities may derive from technical liaisons and technology transfers between commercial products and other goods produced for the government at the government's expense. Externalities may also take the form of more direct financial backup, the simplest form of which is a flow of government procurement that provides enough profit for the company to be able to assume certain commercial risks. The example of the Boeing 747 shows, nevertheless, that a full monopoly position may also create circumstances allowing a manufacturer to recover heavy development costs fully even in the absence of major government externalities.

Hence the market structure, or its regulation, or government externalities, can contribute to creating a sheltered environment favoring monopolistic competition. The innovation process and the organizational pattern of a firm in this kind of environment is that of the second culture. To better characterize it, I shall henceforth use the name *sheltered culture*. In contrast, firms in the first culture sell more standardized products to large numbers of customers. They fight for shares in more open markets, and their organization is generally driven by price competition. To underline their difference from firms in the sheltered culture, I shall henceforth call their organizational pattern and style of innovation the *exposed culture*.

This new definition will enable us to seek out other examples to illustrate and delineate the boundary between the two cultures. In this respect, telecommunications is a particularly interesting field.

Telecommunications: Another Side of the Sheltered Culture

The shape of the telecommunications sector was established in the United States at the beginning of the twentieth century. As the demand for telephones rose rapidly, the American Bell Telephone Company quickly extended its empire. It was achieved via dual integration. The Bell System progressively absorbed independent regional telephone companies. The independents, which carried 51 percent of telephone traffic in 1907, controlled no more than 36 percent by 1921. There was also more and more vertical integration. Aside from local communications (handled by Bell's regional affiliates) and long-distance communications (handled by Bell's affiliate AT&T), the American Bell Telephone Company, via its Western Electric affiliate, manufactured all the equipment needed for expanding the network and meeting customers' requirements. In 1909, after buying Western Union, which carried most of the telegraph traffic, the great American telephone company became the first one in the world able to provide both written and oral communications over its networks.[45] At the beginning of the 1920s AT&T also played a pioneering role in developing radio broadcasting.

The expansion of the Bell System was brought about by several factors. The primary one was the result of positions taken before competitors. Possession of Alexander Graham Bell's patents enabled the American Telephone Company to be the first to negotiate with local communities, under extremely favorable conditions, for setting up telephone networks and services. This gradually created an advantageous source of income for the company, which was already benefiting from economies of scale thanks to its broad geographic scope. A rapid decline in the cost of long-distance communications led to the introduction of equalized tariffs advantageous to residential subscribers. By subsidizing local communications with part of the cash flow generated by its long-distance traffic, the Bell System sustained the expanded demand from new subscribers.[46] Although the dominant position of AT&T bothered a growing number of observers, the intrinsic characteristics of the telecommunications sector led gradually to an acceptance of the idea that this activity should be thought of as a public service based on a private monopoly regulated by the government. This became, de facto, the salient characteristic of the Bell System,[47] and it explains why that system survived for nearly a century in the face of all antitrust actions taken against it. AT&T was only forced to restrain its horizontal integration outside the telephone business. It agreed to sell Western Union in 1913, and to forgo developing its data-processing activities in 1956.[48]

The expansion of the Bell System was also a consequence of a remarkable policy of developing new technologies in the telecommunications sector. From the time of its birth in 1881, AT&T accumulated innovations and technological breakthroughs: the first long-distance link, between Boston and Providence, in 1881; the first transatlantic communication, in 1915; the invention of the coaxial cable, in 1927; the design of the first digital computer, in 1937; the discovery of the transistor principle, in 1947; and the development of Echo I, the first telecommunications satellite, in 1959.[49] Most of these spectacular results originated in Bell Laboratories, AT&T's research branch. Created in 1925, the Bell Labs, as they are familiarly known, are among the most remarkable research instruments ever created. In the 1940s they employed 5,700 people, 2,000 of them scientists. In 1970 they employed 17,000, annually producing some 2,300 scientific articles and 700 patents.[50] Bell Labs publications dominated, in both number and quality, many scientific fields, including solid-state physics, electronics, telecommunications, and radio astronomy. During the year 1956, for example, they provided a quarter of all the research results from industry published in the prestigious journal *Physical Review*.[51] The Bell Labs have produced seven winners of the Nobel Prize in physics.

Bell Labs' organization is unique in blending a large fundamental research component with applied industrial orientations. Notwithstanding the industrial objectives assigned to them, the dominant atmosphere in these laboratories is of great intellectual freedom for the researchers. As one of them pointed out, "Not only does this freedom exist, but it was an essential ingredient in discovering the transistor."[52]

One of Bell Labs' celebrities is BELLE, a chess-playing computer. Beginning in 1973 with a simple program for playing chess on a standard computer, Ken Thomson and one of his physicist colleagues progressively designed a specialized machine. The first model, completed in 1977, included 325 integrated circuits and was able to evaluate 5,000 positions per second. In 1983, when the title of Master was bestowed upon her by the International Chess Federation, BELLE was the world champion chess-playing computer. (In September 1983 she lost her title to a Cray 2, a more recent and powerful supercomputer.)

Thomson, who joined Bell Labs in 1966 at the age of 23, is a brilliant computer specialist. He was one of the inventors of UNIX, one of the most original of all computer operating systems. Combining the use of multiple blocks and programs, the UNIX system provides great user flexibility on a wide range of machines, from micros to mainframes. Thomson is one of the stars of the Bell Labs engineering staff, and the freedom his employers grant him is amply compensated by the benefits they derive from his talent. (In a book entitled *Three Degrees above Zero*, the journalist Jeremy Bernstein reports on the reasons for the success of the Bell Labs.[53])

Enjoying abundant financial resources, and giving their most talented people high salaries and extensive freedom in their work, the Bell Labs provide all the advantages of a university laboratory with virtually none of its constraints. Unlike his university colleagues, a Bell Labs researcher has no need to hunt constantly for new research grants, or to devote a large amount of his time to teaching. This explains the ability these labs have had for fifty years to attract scientific talent. It is a particularly privileged and sheltered world, free of the restrictions and the competition to be found in most industrial and academic laboratories.[54]

The Bell Labs' unique financing system was one of the reasons for this. Their budget came from a levy on the cash flow generated by telephone communications, collected by regional and long-distance affiliates in the Bell System. This method of financing had two advantages. First, it automatically ensured an increase in available research funds as telephone traffic expanded, reinforcing the

Labs' independence. Second, it created a unique pool of scientific resources that was not only available to meet the needs of companies in the Bell System but was open to the outside as well. The Bell Labs have, for more than half a century, constituted a brain pool available to the entire nation. "The reason Bell Labs has been able to maintain this ivory tower atmosphere," says Jeremy Bernstein, "is clearly because their parent firm, AT&T, as a regulated monopoly and one of the richest in the world, had the wherewithal to supply the $1.63 billion that it took to operate the labs [in 1981]."[55]

These favorable conditions for producing innovation and technology are obviously those of the sheltered culture. The discoveries made in the Bell Labs were the result of general progress in basic research, as well as of programmed R&D efforts. It is precisely this emphasis on basic science, in a campus atmosphere, associated with applied research objectives, that made these laboratories so original.[56] For example, in the early 1940s, when limitations in the vacuum tubes then used for long-distance signal amplification became apparent, the Bell Labs began devoting major resources to solid-state physics, which was already perceived as a possible alternative means for amplifying electric signals. This concentration of resources—on the order of $1 million for the three years following the war—made the discovery of the transistor possible a few years later.[57] Bell Labs research has also involved large technological development projects. For example, the first electronic switching system was tested at the Bell Labs in 1960.[58] This example is another illustration of the telecommunications sector as an archetype of the sheltered culture. The cost of developing a switching system today is close to $1 billion. Moreover, it is a highly sophisticated, semi-custom product offering the user multiple options, which make it similar in many ways, with respect to both design and production, to the expensive, complex technological equipment produced in other sectors of the sheltered culture.

Another characteristic that places the telecommunications sector in the sheltered culture is the closeness between its military and its civilian technologies. In this area, as in many other advanced sectors, World War II produced accelerated technical progress. During the war, the Bell Labs were totally devoted to the war effort, and were responsible for half of the research and development work on the first radar units. They participated intensively in military research in the transmission and reception of microwaves,[59] which from 1948 on would be used in the civilian area for transmitting telephone and television signals. Telecommunications-satellite technology also owes a great deal to development work carried out under Pentagon and NASA contracts. More generally, the strategic nature of telecommunications for national defense, and the common wellsprings of and similarities between military and civilian telecommunications equipment, suggest that many ties still exist between civilian and military development in this field, even though it may not be easy to establish precisely their nature and their significance because of the secrecy surrounding military work.

It is probable, nevertheless, that military R&D money makes up only a limited part of the resources invested by AT&T in this activity, which amounted to $2.3 billion in 1986.

The example of telecommunications shows that there are forms of protection other than that provided by the government that enable the sheltered culture to develop and prosper. Unlike the armaments and aeronautics sectors, and excepting the period of World War II, public procurement and direct government financing have played no more than a modest role in the development of civilian telecommunications technologies in the United States. As we have seen, the country's particular historical circumstances led it to create a private quasi-monopoly to supply most telecommunications services and equipment. The existence of this monopoly, as well as certain inherent technical constraints in this sector (such as the important role of networks), has made telecommunications particularly favorable to the development of a sheltered culture. We should note, however—as the experience of many countries has shown—that if telecommunications services have often been developed as a public or private monopoly or near-monopoly, this has generally not been true of equipment supply. Examples of the same company providing both functions are rare. In this regard, AT&T took advantage of an exceptional set of circumstances seldom found elsewhere.

Two major features of a sheltered environment have found their extreme expressions in the Bell System. Here the relationship between the equipment buyers (AT&T's regional affiliates) and the sellers (the Western Electric and Bell Labs affiliates) was taking place within the same company. Hence the importance of the buyer-seller dialogue that shapes the technical characteristics and performance of products, as well as the near-absence of pricing in commercial transactions.

At the same time, the relationship between the Bell Labs and the users strongly influenced the path of telecommunications innovation in the United States. On the one hand, the Bell Labs were remarkably efficient in developing advanced switching and transmission technologies, areas in which their competence was directly usable by the regional companies. On the other hand, they were poorly adapted to generating innovations that could be of direct benefit to telephone users, such as telephone sets and peripheral equipment. Steps taken by Bell in this direction most often resulted in equipment that was inordinately complex or inappropriate and that failed to meet users' expectations. The videotelephone, first tried experimentally as early as 1961, provides a perfect example. The premature development of this device cost AT&T millions of dollars through the end of the 1960s.[60]

This observation highlights the Bell Labs' strengths and weaknesses and contributes to defining the boundary between the two cultures. The development of switching equipment for regional telephone companies and the creation of mass consumer phone terminals derive from two different innovation patterns, from two different types of industrial and commercial organization. The first pattern, centered on the development of custom or semi-custom equipment with unique technical characteristics, is that of the sheltered culture, of which Bell Labs is a pure product. The second pattern, aimed at mass production of low-cost products for large numbers of consumers, is that of the exposed culture. It requires strong marketing capabilities and the ability to anticipate changes in buyers' tastes and needs. It requires above all else a manufacturing organization that minimizes costs. These skills were either poorly developed or nonexistent in the Bell System. This shortcoming was largely responsible for the pressure for deregulation in the telecommunications sector that built up in the 1970s. The situation became increasingly unacceptable as innovative companies outside the Bell System stood ready to offer users new lines of products and services.

AT&T versus IBM

Until it was broken up in 1983 into eight distinct companies, the American telecommunications giant AT&T seemed to possess, at least in appearance, many features in common with the computer giant IBM. First of all, there was size. As the largest US company in terms of revenues ($70 billion in 1983), AT&T outdistanced IBM, whose sales that year amounted to $35 billion. The former employed 950,000, the latter 340,000. Their dominance of their respective markets was the second point in common. AT&T controlled 80 percent of telecommunications services and over 50 percent of the corresponding equipment market, and IBM manufactured nearly half of the computers sold in the United States. Both companies were investing considerable amounts in R&D: $3.5 billion annually by AT&T and $2.5 billion by IBM. During 1983, both companies made sufficient profits to satisfy their stockholders.[61]

Beyond these similarities, however, IBM and AT&T could not have been more different. Technology played a central role in the development strategy of each, but AT&T's new developments were largely determined by available technologies emerging from the fertile Bell Labs system, whereas IBM's reflected a concern for new customer needs.[62] IBM's innovation process was market-driven, and was tuned to satisfying dispersed demand. AT&T's innovation process was technology-driven; its aim was to convince a few customers (regional telephone companies), with whom the manufacturer had a special relationship, that they needed the latest available technologies. Whereas IBM had been managed for many years by sales and marketing executives, AT&T was headed by engineers. It is significant that the position of vice-president for marketing at AT&T was not created until 1973, and that the department responsible for this function was put in place only in 1976. While the telephone company's business enjoyed steady growth, due to the continuous expansion of its captive subscribers' demand, IBM was constantly making a marketing effort and promoting its products in order to keep its market share in the face of increasingly aggressive competition.

The world controlled by AT&T was for many years characterized by a relatively slow rate of innovation and by products with long lifetimes; IBM's world was characterized by a fast pace of innovation and by products with relatively short lifetimes. The result was a centralized management organization on the one hand and a highly decentralized organization and management mode, appropriate for flexibility and adaptation, on the other. In truth, the two American giants are opposites in almost every way. Behind the apparent similarity, each shows the characteristics of the culture it came from: the sheltered culture in the case of AT&T, the exposed culture in the case of IBM.

If we had made the same comparison at the end of the 1960s, however, we would probably not have noted such striking differences between the two firms. At that time, mainframes dominated the computer markets. Even though it was one of the first computer companies to eschew the Pentagon's protection, IBM used the logic of the sheltered culture while expanding its empire during the 1960s. Its technological lead allowed the company to dominate the mainframe market, imposing its latest products on customers. IBM's clientele was captive to the 360 line of products, and IBM's aggressive policy of improvement outstripped the competitors and created customer loyalty. But diversification in the computer field and the falling market share for large systems threatened this domination. For mini- and micro-computers, the competition was no longer solely technological; it was increasingly a matter of price. IBM was among the last of the computer firms to adapt its organization to a market that had already acquired most of the characteristics of the exposed culture. But this had become an absolute necessity if IBM was to maintain its dominant position.

Data-Processing Service Companies: Between the Two Cultures

Software service companies are a little like filling stations for the computer industry—they provide computer users with the indispensable ingredient needed to use the machines. Here a distinction must be made between systems programs and applications programs. The former are tools (operating systems, compilers, database management systems) for a particular type of computer; the latter enable the user to employ his machine effectively to solve a specific problem. Banks, insurance companies, and large and small firms all create different problems for programmers and require different types of applications software. Although the software companies account for only a modest fraction of the market for system programs (half of which are produced by the computer manufacturers), they control a high percentage of applications programs. In the United States this activity has shown the highest growth rate among all industry and service sectors since the early 1980s. In 1987, the total revenues of the software houses, excluding manufacturers' production and software users' internal production, were nearly $80 billion.[63]

An important part of the computer service companies' work is maintenance—updates and improvements of long-available programs. In 1980 this activity generated three-quarters of the American software industry's revenues, with the remaining quarter coming from newly created programs. The reason is simple: most large firms and agencies installed their first computer systems in the 1950s and the 1960s. Unlike other equipment, a program never wears out. Its replacement can therefore be postponed during a period of uncertainty or financial constraint, on the condition that limited improvements enable it to operate with the most recent machines.

The computer manufacturers' policy of compatibility with earlier proprietary software, which provides incentives for users to buy equipment with ever-higher performance, has for many years encouraged this behavior. The Organization for Economic Cooperation and Development estimated the value of programs still in service worldwide in 1982 but written during the 1950s and the 1960s at $500 billion.[64] The computer service companies' daily bread comes from updating and improving software for large users' data-processing systems. Once he has hired one of these companies, the user—whose chief aim is a system that works—has good reason to maintain the relationship. Once a service company is familiar with his firm's data-processing problems, a manager will be just as reluctant to switch to another as he would be to change his family doctor. These captive large-system customers—banks, insurance companies, big industrial firms, and government agencies—provide the major part of the computer service companies' business. Unquestionably, this activity belongs to the sheltered culture: it involves customized work for a small number of customers, and meeting technical specifications is much more important than cost. Moreover, the structure of this part of the computer-services industry reflects this situation. Until recently there was little competition, and labor productivity was low.[65] Aside from affiliates of major computer manufacturers, the large American computer-services companies include many affiliates of Pentagon contractors, such as TRW, General Electric, and McDonnell Douglas. Military programs carry significant weight in this industry.

Another side to this industry appeared in the late 1970s with the technological revolution of computer networking. This marked the beginning of the decline of centralized computer systems, which were replaced by smaller machines linked together in networks. For professionals in computer services, this change meant new opportunities, particularly in standardized software. New software packages began to be mass-produced for large numbers of users, bringing a revolution to the world of data-processing services. In a sense it was like introducing ready-to-wear in a profession that until then had known only high fashion. The manufacturers of personal computers were the first to jump on the bandwagon and actively engage in the battle for software packages. Developing systems and applications software was in fact the surest way to get more people to use microcomputers.

But ready-to-wear is very different from high fashion. Manufacturing software packages is a high-risk activity requiring large investments (developing new standardized software may take several years) for an uncertain outcome. Marketing software also involves expensive promotional campaigns. For example, the introduction of 1-2-3, in 1983, cost Lotus nearly $3 million. Lotus was aiming for mass production, in order to be able to lower unit prices, so as to cover the development costs. That strategy succeeded, and within a few years Lotus was number one in its field. In 1986 Lotus' gross sales amounted to $283 million. Jim Lanzi, president of the company since 1984, was the mastermind of this success. He was a marketing man who gave Lotus a commercial infrastructure unlike that of any other company in its sector.[66]

A different approach was taken by Bill Gates, the young president of Microsoft, Lotus' principal competitor. Gates is a specialist at making "killings" in the microcomputer industry. He was able to develop the software that was adopted by the major computer manufacturers. One of his most spectacular successes was his 1980 sale to IBM of Ms-DOS, the operating system used for IBM's first personal computer. Used not only by owners of IBM PCs but also by most owners of IBM-compatible machines, Ms-DOS is the top-selling personal computer software in the world. Recently, by assigning Microsoft the task of designing the operating system for its second-generation personal computer, the PS, IBM positioned Microsoft as the industry's software standard setter.[67]

Lotus and Microsoft are running nose to nose in several segments of the software-package market. Competition is harsh in this sector, stimulated by the arrival of many new firms. Software packages have become the terrain of choice for new entrepreneurs and venture capitalists.

Here we are far from the traditional realm of large computer-service firms such as TRW, Computer Science Corporation, and Automatic Data Processing, which are doing business worth more than $1 billion a year in custom-made software but which would probably be incapable of competing for market shares with Microsoft, Borland, or Ashton Tate,[68] the rising stars of software packaging. As we have seen, it is not easy to pass from the logic of the sheltered culture to that of the exposed one.

By the mid-1980s, the data-processing services sector had already undergone a profound transformation. Two distinct professions were emerging, with two different objectives and organizational modes. At the interface between these two worlds, firms with different logics were competing for market shares with increasing bitterness. But the sheltered turf of the traditional computer service companies was being increasingly threatened by the fast-growing software houses in the exposed sector.

The Pharmaceutical Industry: Another Model of the Sheltered Culture

In ten years, Roy Vagelos made Merck one of the most prosperous companies in the American drug industry. A brilliant biologist at the National Institutes of Health and then at Washington University in St. Louis, he joined Merck in 1976 as research director. He became chief executive officer in 1985. In the meantime he turned the firm into one of the most innovative in its field. During 1986 alone, the company introduced five new drugs.[69] This was due to a vigorous R&D effort, in which the firm invested $460 million a year—12 percent of its revenues.[70] Resources must be provided for research into a large number of new molecules in order to market even a few. Bringing a new drug to market is like running a steeplechase. Once a new molecule has successfully passed the first barrier—having been shown to possess useful pharmacological properties—there is still a long road ahead. The second stage, which may last from one to three years, involves exploratory development. Here the aim is to synthesize the new compound in large quantities, and make preliminary tests to see how it is absorbed and tolerated by the body. The third stage, generally lasting a year, involves studying the new molecule's efficaciousness and toxicity through limited clinical tests. The fourth stage, lasting two to five years, is that of real-life testing, which may involve as many as 2,000 patients.[71] The results obtained during this stage determine whether the final barrier—getting the Food and Drug Administration's approval to market the product—will be crossed.[72] As the CEO of a large pharmaceutical company recently commented: "Developing a new drug is like betting on horses, with one notable difference. At the track you at least know one horse will end up a winner. But there's no guarantee you'll have a winner with drugs."[73]

Although it is a high-risk activity, the production and sale of drugs nevertheless can bring great rewards to those with the requisite resources and patience. One of Merck's greatest successes in recent years was Vasotec, a drug for hypertension. Vasotec's sales exceeded $250 million in 1986 alone; its total development cost had been of $100 million.[74]

Once it has obtained FDA approval to market a product, a pharmaceutical firm will generally be eager to recover its R&D investment by pricing the new product as high as possible. In some cases, a product with unique properties that meets an urgent need may be sold at an astronomical price that is only partly justified by its complexity and its production cost. Patent protection contributes to such circumstantial advantages. At the same time, patent protection is legitimate, for this is what enables the company to recover its initial R&D investment after running the obstacle course to the market. One of the most extreme recent examples is AZT, a drug used to treat AIDS. Burroughs Welcome introduced it to the market at the price of $10,000 per year per patient.

Aside from the unique characters of certain drugs, which make them rare commodities for which users are prepared to pay dearly, the economic logic specific to the pharmaceutical industry stems in part from the fact that the true user of the drug is not the patient but the physician prescribing it. But the physician does not have a true commercial relationship with the drug's manufacturer. His main concern is not the drug's price but its technical characteristics. In its logic, the pharmaceutical industry's operating mode is truly that of the sheltered culture. The shelter that enables this logic to prevail is the government's regulation of the drug market as a means of protecting public health. A large part of the drug companies' energy and resources is devoted to dealing with regulatory procedures, which serve as significant protection against newcomers in the market. As Merck's general manager has observed: "You can build a computer in a garage. You can have a great idea for a drug. But to get the ultimate molecule takes enormous effort, and it's not going to be done in a garage."[75]

Americans bought $12.6 billion worth of drugs in 1983, of which only 6 percent were imported. This figure illustrates how hard it is for a foreign firm to sell its products in the United States unless they are manufactured there. But foreign companies that are tempted to make drugs in the United States will have to face one of the world's most complex regulatory systems, and one of the slowest to grant licenses.[76]

However, there is another aspect of the pharmaceutical industry that largely escapes these constraints. This is the over-the-counter drug market. Intense price competition has developed in this sector, which represents slightly less than 40 percent of American pharmaceutical sales. The logic of the exposed culture is dominant in this market.[77]

The Territories of the Two Cultures

The preceding examples have progressively highlighted the differences between the two cultures. We can now see more clearly that the boundary between them runs through most sectors of high technology. However, in most of the sectors, one of the two logics is predominant. Remember, this distinction is not only of intellectual interest; it reflects two different types of management and industrial organization, two different categories of firms. Furthermore, it is difficult for a firm to move from one category to the other. Consequently, knowing the extent to which each of the two cultures influences each industrial sector can be extremely useful for understanding that industry's development pattern and evaluating its capacity to withstand international competition.

The time has come, therefore, to survey briefly the different high-technology sectors, and to attempt to characterize them with respect to the two cultures. Before doing so, however, let me spell out once again the criteria that will be used to draw the dividing line.

The initial hypothesis underlying the distinction between the two cultures is that there are two different processes of innovation, associated with two different types of company organization. This is the hypothesis that led us to examine the nature of the relationship between seller and buyer when high-technology products are involved. The nature of this relationship should provide a key— our second hypothesis—enabling us to determine to which culture the selling firm belongs. We saw above that, in this regard, two different situations can arise.

In the first situation, price is less important for the buyer than the satisfaction of technical requirements. Availability of technology or know-how is the principal criterion for purchasing, and in some cases it may be the only one. In such a situation the supplier tends to adopt monopolistic behavior. He maximizes revenues by playing on the relative scarcity of his product and on its technical advantages over competitors. Here, competition is technology-driven. We use it as our principal criterion for determining whether a firm belongs to the sheltered culture. It is important to emphasize that this competitive mode does not necessarily mean an absence of competition, even if it seems insignificantly weak in some sheltered markets. "Sheltered culture" is not necessarily synonymous with "defense industry." In this sheltered universe we find a large proportion of firms that have proved able to stand at the forefront of technological development. The structure of demand often corresponds to a small number of technically knowledgeable customers buying small numbers of custom or semi-custom products. This explains why the innovation process here is focused on the products rather than on the production process. Moreover, this logic often leads to technological breakthroughs rather than incremental innovations. The sheltered culture is, to a certain extent, the high fashion of high tech.

In contrast to this are the situations in which the key to winning or maintaining a market share depends on a firm's ability to lower its production costs. To use the language of economics, we would say that demand here is highly price-elastic, whereas in the first situation it is not. This means that the lower the price, the more the manufacturer will be able to sell. This situation characterizes the exposed culture. It usually corresponds to the position of a firm wishing to reach a dispersed clientele with standardized products. Note that the importance of the price criterion does not prevent the firm from using different product lines to reach different market segments with differentiated or personalized products. But the logic of market penetration is the same, and—in contrast to the sheltered culture—its aim is to reach a broad customer base. Under these conditions, process innovation is a key to success. Moreover, its focus is frequently on incremental improvements rather than on fundamental changes. The exposed culture is thus, in a sense, the ready-to-wear of high tech.

Unlike the exposed culture, the sheltered culture may have several aspects. It cannot develop unless the company has the benefit of a shelter or some form of protection limiting competitors' access to the market. But such protection may be of different kinds: it may exist because the buyer is the government, as in space or armaments; it may result from a government-regulated market allowing high profit margins, such as in pharmaceuticals; or it may be the consequence of a regulated near-monopoly, as in telecommunications; or it may result from certain characteristics of the technology itself. For example, the need to own or control access to networks constitutes a serious barrier to newcomers in the tele-communications industry. Some technical constraints may also tie users to their suppliers. Custom-made computer software falls, as we have seen, into this category. This is also true of many complex or specialized products; in these cases, the possession of unusual know-how or technology and a long-standing relationship between buyer and seller serve as barriers to competition. For various reasons the customers become captive. Numerous items for space, aero-nautics, and nuclear energy belong to this group, as well as many advanced materials, sophisticated medical imaging equipment, and large-scale custom-made computer software. Most of the compa-nies manufacturing these items enjoy the protection of what is usually called a "technological niche," except that in this case the protection is permanent. Classically, a technological niche is, in effect, the position of a firm with a unique product, the manufacture of which is controlled by a patent or by know-how that the firm alone possesses. The larger the corresponding market, the faster the company is likely to be dislodged from its niche, or in any case to be required to adapt to the arrival of new competitors.

As early as 1939 the American economist Joseph Schumpeter[78] had noticed that the usual rules of competition were not applicable in certain high-tech sectors. Continual investment in the creation of superior, differentiated products for which the innovator holds a temporary monopoly distinguishes these sectors from more mature industries in which products have reached a higher level of standardization. The existence of a monopoly, and the situational advantage associated with it, is in fact the very thing that enables a company to accumulate sufficient profits to continue its investments in R&D. Raymond Vernon, in his theory of product cycles,[79] observed that this kind of situational advantage coupled with the filling of technological niches is much more frequent in the initial phases of technological development. The newer an industrial sec-tor, the more niches there are. Growing to maturity limits their number.

The models that attempt to explain the behavior of firms in high-technology markets have since become more sophisticated. For example, it is widely accepted that continual technical progress is to be found over fairly long periods in certain sectors, allowing firms to perpetrate the behavior noted by Vernon and Schumpeter

Table 2
The major high-technology markets in the United States, from the most
sheltered to the least sheltered (in the early 1980s).

	Percent sheltered
Armaments and space	95
Nuclear and large electrical equipment	90
Advanced materials	80
Aeronautics	75
Professional electronics	75
Telecommunications	70
Pharmaceuticals	65
Software	60
Data processing	30
Electronic components	20
Office equipment	15
Consumer electronics	10

by moving from one product to another. This observation applies
fairly well to many companies belonging to the sheltered culture.
However, by our definition of this sheltered culture there are other,
nontechnical factors—and this is an important point for character-
izing our two-culture model—such as the demand structure or
market regulations, which enable firms to perpetrate the same
behavior over a very long period. Together, these characteristics
now help us identify which of the two cultures a particular firm
belongs to, because firms able to defend themselves on both terrains
are rare indeed.

Table 2 represents, for each of the major high-technology sectors, a quantified estimate of the influence of each of the two
cultures. The purpose of this table is only to show orders of magnitude of the sheltered and exposed markets. The boundary sepa-
rating them is necessarily fuzzy, given the multitude of intermediary
cases. In fact, it is really a continuum we are dealing with, with the
most exposed firms and markets at one end of the spectrum and
the most sheltered ones at the other.

A reading of table 2 suggests that companies struggling for
market shares in Silicon Valley or along Route 128—the archetypes
of the exposed culture—are only the visible tip of the high-tech-

nology iceberg in America. The entrepreneurial and highly competitive vision of high-tech industries does not reflect the most frequent situation; it reflects only a small part of these markets. Difficult-to-reach markets, in which a smaller number of firms are often enjoying an oligopolistic or monopolistic position, following another logic and other rules for success, are more prevalent. Most of these firms have the asset of a shelter against the intrusion of outsiders.

Technology and Politics

5

Like any other director of an important federal agency, James Wyngaarden of the National Institutes of Health spends a great amount of time defending his organization's needs before those responsible for drawing up the annual federal budget. For Wyngaarden, however, this perilous task is exceptionally easy, as we shall see. A look at the process used to fund the NIH reveals a great deal about the relationship between the executive and legislative branches.

A Pluralistic Decision-Making System

The budget process in the United States begins in the fall and lasts throughout the year. The fiscal year starts on October 1, at which time each of the federal agencies receives its new yearly endowment. The first phase of the budget process happens within the administration. For research agencies it involves routine discussions between the head of the agency and the department of the executive branch to which he reports—in the case of the NIH, the Office of Science and Technology Policy—along with the powerful Office of Management and Budget. This first phase ends in late December, when the budget forecasts for government research agencies are integrated into the thick file that describes the government's proposals for the overall federal budget, in which the last tradeoffs will be made at the White House in early January. The major points in this budget project will become part of the president's State of the

Union speech at the end of January. Over the following weeks, each of the department and agency heads will explain details of the proposal to specialized committees in both houses of Congress. Among them will be the president's Adviser for Science and Technology, who is the director of the OSTP and whose task it is to present budget proposals for the funding of research.[1]

In Congress, scientific questions are generally handled by the House Science and Technology Committee and the Senate Commerce, Science, and Transport Committee. However, since research activities are spread out over many government departments, some questions touching on this area can also be handled by one of the 22 other committees in the House or the 16 others in the Senate.[2] There is no overriding authority for the science an technology budget in Congress. Budget tradeoffs are made, within each committee or subcommittee, between scientific projects and other programs that may cover questions as different as health insurance for the elderly and welfare.

Some 200 subcommittees, in both houses, are responsible for examining the government's annual budget proposals. The outcome of budget discussions in a given committee will depend on which political party controls the committee. The greater the opposition's influence, the greater the chances that a government proposal will be amended. The worst case for the government is one in which the president's party does not control either of the houses, which has been true since the 1986 elections. Relations between the legislative and executive branches have become severely strained, and the president's power has been weakened. The constitution provides for a subtle balance of power between Congress and the White House. For most important decisions, and financial ones in particular, it is the president who makes proposals and the Congress that accepts or rejects them. For this reason, continual negotiation is part and parcel of the relationship between these two branches.

Since 1985 there has been a hardening of the respective positions, with Congress becoming increasingly resistant to the administration's proposals for legislation.[3] Under such circumstances only two options are available to the president. The first is to negotiate from a position of weakness with a hostile Congress, which often means that he is forced to accept a gutting of his initial proposals. The second is to veto bills passed by Congress. In the case of a presidential veto, Congress must reexamine the bill but may decide not to change it. In this case a two-thirds majority in each house is needed to override the veto. Since the Democrats cannot muster this percentage in either house, the veto should theoretically serve to rescue many projects. In practice, it doesn't work this way. Whereas the members of a political party in most other Western democracies will act together in parliamentary voting, this is not always true in the United States. In recent years Republicans have often joined Democrats to approve bills opposing various government proposals. This has happened frequently with respect to the defense budget. It has also happened to the NIH budget.

As was mentioned above, NIH Director James Wyngaarden has the job of defending his budget proposal before the House and Senate committees that deal with scientific matters. Once he has explained his budget request, he can usually expect to be told by the committee that he should have asked for several hundred million dollars more. The president will then predictably veto what he sees as unacceptable congressional generosity toward the NIH. Just as predictably, there will be enough votes in Congress to override the veto and confirm Congress' counterproposals. This is how the NIH came to have a 1988 budget of $6.67 billion—a 7.7 percent increase over 1987—when the White House had wanted to limit it to only $5.61 billion, a 9 percent drop from the previous year.[4] Here, however, the president's defeat has no special political meaning, as it does in the case of other matters such as the defense budget. It simply illustrates the powerful influence with Congress of the medical research lobby. The strained relationship between Congress and the administration over the NIH budget is nothing new. In 1959 President Dwight D. Eisenhower was already vetoing congressional generosity toward medical research, and losing. At that time the NIH budget was $547 million, 37 percent more than the executive branch had asked for.[5] This example illustrates the complex power sharing involved in budget decisions.

The US government allocates some $130 billion per year to research and development,[6] for a multitude of purposes including national security, public health, industrial competitiveness, scientific and technical training, and the advancement of knowledge. In order to understand the logic of these allocations, we must first understand how pressure groups—lobbies—are formed to further the interests of certain groups within the scientific and technological establishment, and how they wield influence.

An Honorable Cause

Shortly after World War II a number of prominent persons played important roles in orienting part of the federal R&D effort toward medical research. Mary Lasker was one of the most influential in this respect. In 1942, with her husband Alfred, she created the Lasker Foundation and began giving awards to outstanding researchers in medicine. Today the Lasker Award is still a coveted distinction, one that frequently precedes the Nobel Prize in medicine.[7]

Alfred Lasker made a fortune in advertising, and Mary Lasker devoted herself to charity, especially during the war. While working for several private health institutions, she came to the conclusion that nothing significant could be done in this field without federal assistance. This opinion was shared by her friend Florence Mahoney, whose husband was the manager of the *Miami Daily News*, which progressively became an advocate of medical research. The Laskers and the Mahoneys were friendly with Senator Claude Pepper of Florida, who was himself interested in health matters. When the Laskers visited the Mahoneys in Miami during the spring of 1944, they made two decisions. The first was to campaign for Senator Pepper's reelection, with the Laskers providing financing and the Mahoneys the columns of the *Miami Daily News*. The second was to ask the senator to use his influence in Congress in favor of medical research.[8] Pepper was reelected, and he shortly organized public hearings on the subject. Sixteen experts, including several prominent persons recommended by Mary Lasker, testified. The hearings served to convince a certain number of congressmen of the need for the federal government to help medical research, which until then had been financed exclusively by private foundations. At the same time Mary Lasker started a subscription campaign for cancer research and, with her husband, soon took leadership of the American Society for the Control of Cancer. In 1945 the Laskers collected $4 million in endowments, five times as much as the society had collected before their arrival on the scene.

In 1946 they collected $10 million. In 1970 the foundation, which in the meantime had become the American Cancer Society, collected $70 million.

The Laskers' connections in Congress diversified rapidly. After the 1946 elections, Senator Henry Styles Bridges, a Republican who replaced the Democrat Pepper as chairman of the senate committee in charge of medical affairs, became a close friend of the Laskers; he benefited from their help toward his 1948 reelection. That year Bridges helped persuade President Harry Truman to create the Lasker-recommended Heart Research Institute.[9]

The foundation was now in place for a large, government-financed medical research center. From $7 million in 1947, the NIH budget reached $70 million in 1952. "What was now needed," wrote Stephen Strickland in his book on the American medical lobby, "was someone in Washington watching over the medical research scene full time, scouting for potential allies, reinforcing existing support, keeping present friends happy, and especially, coordinating the annual approach to Congress for greater outlays for the Cause."[10]

The man for the job was Mike Gorman. A medical reporter, he was discovered by Florence Mahoney. Through her efforts he soon became executive director of the President's Commission on the Health Needs of the Nation. The commission's report, written by Gorman, recommended the establishment of a national committee on mental health. It was created in 1953, with Mary Lasker as president. Freed from his government obligations, and conversant with the ways of Washington, Mike Gorman became the first lobbyist for medical research.

By the mid-1960s Gorman was on first-name terms with 150 members of the House of Representatives. He could also directly call a few senators. For congressmen he was a precious source of information on medical matters. He was able, on short notice, to locate competent and reliable experts on a wide variety of subjects. Stephen Strickland gives several examples illustrating the lobbyist's role: "In 1967, Mike Gorman persuaded Senate Majority Leader Mike Mansfield of Montana to host a luncheon for a number of his colleagues so that they could be brought up to date on the latest advances, and projected needs, in the field of heart research. Thirty-six senators attended, and heard Dr. Michael DeBakey, the heart surgery and heart transplant pioneer, and other distinguished cardiologists describe the state of heart medicine. The doctors also reminded the senators that one million of their fellow citizens die of heart disease each year. Gorman saw to it that the meal was edible and that the tables were bussed properly. Senator Mansfield made opening, sympathetic remarks. Mrs. Lasker paid the check."[11]

Mike Gorman's situation is, of course, not unusual, and his calling is perfectly respectable. Any individual, company, or special-interest group concerned with promoting an idea or a project in the federal government, with seeing a bill passed in Congress, or with defending the interests of a group or a geographic area, or seeking access to various government contracts, may make use of lobbyists to identify and exploit the most effective channels for communicating their views through the labyrinth of the government decision-making process.[12] Lobbyists and specialized consulting firms providing similar services number in the tens of thousands.

Pork-Barrel Politics

Lobbying is an essential ingredient of the American sheltered culture. In this world, as we have seen, the projects the government will choose to finance are based on proposals made by scientists, industrial firms, and federal research agencies. Here lobbying plays an important role in influencing government decisions.

By the end of the 1950s lobbying for specific scientific and technological projects had become widespread, and it gained in importance as the government began launching ambitious programs in space exploration, transportation, nuclear power, supersonic aircraft, and so forth. Lobbying is usually aimed at particular committees in each of the houses of Congress. It may be backed by a federal agency, such as NASA, the Atomic Energy Commission, or the Federal Aeronautics Administration, that is itself involved in the project. For many years the nuclear-power lobby was one of the most efficient. It had the unusual privilege of having to address only one committee for both houses of Congress: the Joint Committee on Atomic Energy, which was under the control of a few influential congressmen and senators who were favorable to the cause.[13] It played a particularly important role through its successful backing for the controversial Clinch River breeder reactor project, which the government had long wanted to abandon. This demonstrates the extraordinary influence certain pressure groups may wield within the sheltered culture.

Lobbying is particularly intense when there is a government project in search of a geographic site. In such cases, rivalries and competitive bidding occur between states, and pressure groups involve themselves in "pork-barrel politics" trying to attract public money to a project. A typical example is provided by the case of Fermilab. The 1963 decision to build a billion-dollar proton accelerator led to innumerable deals and battles for influence in Congress over where the machine would be built. President Johnson decided in 1967 that it would be located in Batavia, Illinois.[14] Similar lobbying began in 1987 when it was announced by President Reagan that a decision had been made to build the Superconducting Supercollider—a giant proton accelerator with a 60-mile circumference, 20 times as powerful as the one at the Fermilab. Twenty-five states lined up to solicit this project, whose cost was estimated at $5 billion and whose "fallout" would represent 2,500 permanent jobs, plus an additional 4,500 during the construction phase, for the area where it would be located. Such high stakes explain why a good number of states created organizations to study how best to satisfy the project's requirements, and to promote their cause in Congress.[15]

Pork barrel also plays a role in federal funding for universities, and in defense contracts. Such interregional warfare is not peculiarly American. Struggles for influence underlie governmental decisions in most parliamentary democracies. In the United States, however, institutionalized lobbying suggests that the problem is handled more openly and less hypocritically than in Western Europe. This is a result of different political traditions and government organization. It is also a result of the legislature's playing a more important role in public decisions in the United States than in France, for example. Institutional lobbying is no doubt part of the price to be paid for a true balance of power in a pluralistic society in which the opinions of individuals or groups of individuals can be taken into account and respected—provided they can afford a lobbyist!

Federal Power and Local Freedom

The average American mistrusts the federal government as a matter of principle. This is a visceral mistrust anchored in history. Many early immigrants were fleeing from political oppression or religious discrimination in Europe. Once the new nation was freed from British rule, a chief concern of the various ethnic groups in the thirteen original states was to make it impossible for one of them to seize control of an overly powerful central government. The primary object of the first legislators was, at all costs, to preserve local freedom from being usurped by the state. The most important dimension of this local freedom was economic. The right to engage in one's own enterprise and to create one's own wealth was one of the young nation's fundamental principles. This principle was easy to apply inasmuch as America, at least through the nineteenth century, was not subject to one of the biggest constraints facing governments: the allocation of land and natural resources available within narrow geographic boundaries. The westward-moving frontier gave credence to the idea that America had unlimited resources, and that all one needed to do was to exploit them in order to become rich and contribute to national prosperity. At the same time, the apparently limitless territories yet to be conquered constituted a safety valve for relieving the tensions and social conflicts arising in the industrialized cities of the East.[16] Under these conditions the role of the federal government was limited to a few essential functions such as national defense, postal services, and immigration. In 1870 the entire government apparatus employed fewer than 50,000 persons, of whom two-thirds worked for the Post Office.[17]

Toward the end of the nineteenth century, the industrial landscape changed with the development of mass production. By 1880 the unprecedented increase in productivity had led to severe overproduction in many industries. The creation of large cartels prevented the collapse of prices but aroused public suspicion of industry. Moreover, the cartels did not do anything to alleviate the social cost of industrialization. At a time when European workers were pressing the demands that would transform industrial organization in the Old World, American trade unions still had very little influence, and the federal government was practically uninvolved in social issues. "Poverty and disease, no matter how widespread, were still considered matters of personal misfortune rather than manifestations of social dysfunction. The appropriate response, therefore, was assumed to be private charity rather than public programs."[18]

Growing social problems, linked with rapid urbanization, poor working conditions in factories, and new waves of starving immigrants crowding into ghettos, contrasted sharply with the financial and industrial fortunes amassed by Rockefeller, Carnegie, and Vanderbilt, among others. The wave of anti-business sentiment that shook America at the end of the century led to the Sherman Antitrust Law of 1890 and to the first government regulations in labor and social matters.

Even more basic was the appearance of a cleavage between two opposing schools of thought that would emerge in American society, remaining even to this day. The "business ethic" extols the virtues of a free-market economy as a means of solving society's problems and considers business to be vested with the essential mission of ensuring America's prosperity; the proponents of the "civic ethic" insist on the incapacity of business to resolve the growing problems facing American society, and call upon the federal government to solve them.

Paradoxically, both camps trace their values back to the very roots of American society. The business-ethic camp praises individual initiative and entrepreneurial freedom, which were the outstanding values of the early pioneers, while the civic-ethic camp insists on the citizen's role within the group, the community, and the city, and on joint responsibility for group survival and individual success.

"In countless ways," says economist Robert Reich, "Americans are called upon to choose between these two sets of central values—social justice or prosperity; government or free market; community or freedom. A debate over environmental pollution becomes a contest between the vision of a restored community flourishing within a scenic and healthy environment and the opposite vision of unfettered entrepreneurs whose ambition and daring would create new products and processes to benefit all."[19] For the past century, says Reich, a pendulum swing between these values has alternately brought to power men representing each.

After the first wave of regulation, at the turn of the century, the proponents of the business ethic regained influence during the decade following the First World War. When they showed themselves unable to solve the problems raised by the Depression, they gave way to the New Deal interventionists. They returned to power after the Second World War, however. Through the mid-1960s their vision coincided with the values extant during the most prosperous period America had ever known. But by the late 1960s their influence had declined, to the benefit of civic-ethic proponents brought to power by growing demands to protect the environment. They, in turn, were victims of the economic crisis that in 1980 brought Ronald Reagan and the business ethic to power.

This alternation often, although not always, coincides with the alternate ascendance to power of the two great American political parties. The Democrats are usually in command during periods of civic activism, when a consensus exists for government intervention. The Republicans are usually brought to power by a wave of economic liberalism, when values swing toward free enterprise.

Underlying these changes is a constant distribution of tasks in American society: it is the government's job to promote social justice in accordance with the values expressed by public opinion, and it is the job of business to provide prosperity. This is one subject on which most Americans can agree.[20]

Federal Involvement: From Principles to Reality

From this sharing of responsibilities springs a rule that defines relationships between the federal government and business: the principle of government neutrality. This principle, to which both Republicans and Democrats hold, states that government's involvement in business and industry should be minimal as long as market mechanisms are effective and the rules of free competition are guaranteed.

An enormous gulf may sometimes stand between this principle and reality. Over the years, the tenet of government neutrality has been subject to many affronts, but this has not prevented the dogma from remaining as vital as ever. Each decision that government involvement was required by the needs of the moment—for example, preventing the bankruptcy of the Chrysler Corporation in 1981—has been followed by a tendency to consider such decisions as an exception and not as a rule.[21] The first step in the process that has, *de facto*, led the government to play an important and increasing role in industry and the economy was taken during the First World War. "The experience of the war and of the federal government's role in managing the nation for war legitimized in many people's eyes the role of the government (and, in particular, of the president) as general manager of the economy, responsible for the overall health of American business."[22]

In the 1920s, the rationale for government involvement in industry was to lessen the impact of economic cycles and avoid the dangers of overproduction. For this purpose, federal agencies were made responsible for establishing a framework in various industrial sectors, in close symbiosis with industrial associations (the creation of which was encouraged). The growing influence of these associations and their close relationship with government bodies led to the emergence of a superstructure of industry-wide management, which was effectively in charge of coordinating industrial investments and major production decisions. This superstructure played a decisive role in regulating several large industrial sectors, including automobiles, steel, energy, and transportation, and contributed to the stability of the organization of industry in these sectors.[23]

This system was little affected by political changes. If Franklin D. Roosevelt's rise to power in 1933 meant greater government involvement in the economy, this involvement only confirmed policies for regulating and coordinating industrial production that had been implemented during the 1920s. Officially the federal government's role was limited to overall coordination and regulation to guarantee free competition. Between the wars, a series of antitrust suits reminded industrialists that there were laws with respect to industrial concentration and organization. In reality, the relationship between government and industry was ruled by the powerful industrial associations, generally controlled by the largest firms in each branch and defending the interests of those firms in government circles and federal agencies. Policies implemented in each sector directly reflected these influences, and where backed in Congress by the representatives from the states where these industries were established. The steel sector provides a prime example of this type of relationship between the federal government and industry. "The first chairman of US Steel, Judge Elbert H. Gary, became de facto manager of the industry. He even hosted dinners for steel investment and production. Gary also worked closely with the federal government. . . . Gary's understanding with the federal government inaugurated fifty years of stability for the steel industry."[24]

This special relationship also enabled the American steel industry to obtain from the government protective barriers against foreign competition. In fact, beginning in the 1960s, this concern turned out to be the principal justification for the federal government's involvement in American industry, especially in favor of declining industrial sectors. This involvement took various forms, but chiefly it involved tax relief and low-interest long-term loans. In 1950, tax breaks for the industry amounted to $7.9 billion, and preferential interest rates were costing the federal government $300 billion. In 1980, these figures were respectively $62.4 billion and $3.6 billion.[25] Protection was also provided via a complex system of customs duties and quotas. Industry associations were continually calling the government's attention to the "unfair trade practices" of their foreign competitors, accusing them of "dumping" products at prices below cost. Such claims were often the starting point for the negotiation of agreements between the federal government and the governments of the countries involved, to set "voluntary limitations" on their exports to the United States. Such was the case for consumer electronics and steel in 1977, and for Japanese electronics components in 1986. Between 1972 and 1977, the Commerce Department held 122 investigations of dumping at the request of various industry associations, 34 of which led to import restrictions.[26]

Other kinds of government involvement predominated in high-technology industries, and in other growth sectors. They fall into two major categories.

The first of these other kinds of government involvement occurred in sectors whose expansion involved infrastructure equipment for transportation, communication, and the generation and distribution of electricity, and whose operation was regulated in the public interest. Such regulation often resulted from decisions made at the federal and the state levels. For example, in the telecommunications field the need for regulation emerged in the late nineteenth century, when American Bell Telephone was extending its empire and its quasi-monopoly as the provider of telephone services over a growing portion of the United States. In 1881, after many complaints against the phone company, Indiana became one of the first states to pass laws fixing local telephone rates.[27] In 1934 Congress enacted the Communication Act, creating the Federal Communications Commission to oversee existing federal regulations. The FCC had the power to set long-distance telephone rates and amortization rules for equipment, as well as to determine the conditions allowing competing networks to be interconnected. Its decisions were implemented at the local level by the Public Utility Commissions. In 1938 the new agency brought suit against Western Electric, the American Bell Telephone affiliate that held a quasi-monopoly in the manufacturing of telephone equipment. Western Electric was accused of charging local companies exorbitant prices, and thus contributing to rising telephone rates. In the 1950s the FCC implemented equalized tariffs, allowing subsidies for local telephone traffic, thanks to the declining cost of long-distance communications, the volume of which was rapidly increasing.[28] But in the early 1970s, this regulatory system, which had been created progressively, became subject to growing consumer criticism, because it was unwieldy and because it reinforced the dominant position of the Bell System. Deregulation had become the order of the day, and new rules were set up to stimulate competition between suppliers of equipment and services.

In air transportation, the rationale and circumstances of government intervention were different. The creation by local communities and by the federal administration of airports and navigational aids had great impact on the expansion of air traffic. Between the wars, public subsidies for mail transportation—the chief source of air traffic—contributed to the development of airline companies. In those days, the airlines were often affiliates of aircraft manufacturers (United Airlines, for example, was a Boeing affiliate). In 1934 the Mail Act put an end to this situation. It also gave the Interstate Commerce Commission the responsibility for setting air-transportation tariffs. In 1938, the Civil Aeronautics Act created a regulatory framework based on the principle of limited competition between airlines. The Civil Aeronautics Board was given responsibility for licensing airlines to operate, with the condition that the number of companies on any given route be proportional to the traffic. This regulatory framework was to remain in effect (albeit with some modifications) until 1978, when air transportation was deregulated.[29]

In these regulated sectors, government action—even when indirect—had a crucial influence on the organization of the industry.

The second type of government intervention in high-technology sectors is the result of the unique role played since World War II by technological development in an area that had traditionally been the responsibility of the federal government: national security.

The Federal Government and Technological Development

For the proponents of the business ethic, technological development was only a pretext that would allow a new technocratic elite to come to power and to extend government intervention, thus distorting American values.[30] The irony is that it was under a conservative Republican administration, led by Dwight D. Eisenhower, that the first important decisions on government involvement in technology (involving, for example, the creation of NASA and the accelerated development of ICBMs) giving birth to this "technocracy" occurred.[31] These decisions were seconded by John F. Kennedy, who made available the resources for launching what a few years later would be seen as the most extraordinary military and civilian technological effort ever seen. Under Kennedy there arose a strange alliance between Pentagon "hawks," who dramatized Soviet progress in order to obtain greater military R&D funding, and liberal Democrats, who saw in these expanded programs a way to extend the federal government's influence.[32] During this period, which lasted through the mid-1960s, the government also initiated large technological programs in other civilian sectors. The nuclear-power program, started after the war, and the supersonic-transport program, approved by Kennedy, illustrate the growing role of government in technological development.

During the subsequent period, lasting through the 1970s, these orientations were called into question for several reasons. For one thing, the Republicans, who regained the White House with the 1968 election, were openly skeptical about the large government programs launched under the two previous Democratic administrations; their feelings reflected those of a growing segment of public opinion. Second, the Vietnam war revealed the destructive power of new weapons produced by technological progress, such as defoliants and laser-guided bombs. Moreover, the ecology movement was gaining influence.[33] The supersonic transport, abandoned by President Richard Nixon in 1970, was the first victim of the new but already powerful environmental lobby. It would be only another five years before the nuclear-energy program was also killed.[34] The emerging ambition was to reorient science in order to resolve society's real problems. This was reflected in Nixon's 1971 cancer-research program. Five years later, Jimmy Carter rallied the scientific and industrial community around a new national priority: energy independence.[35] Underlying these renewals of R&D priorities was a constant belief, characteristic of government intervention: regardless of its new target, the government assumed it was enough to define the goal and apply science and technology to reach it. With some variations, government intervention still followed the approach of the Apollo era.[36]

This reasoning seemed to change in the late 1970s. The recession that followed the 1974 oil crunch brought to the fore a concern that had been latent since the beginning of the decade: American industry's loss of competitiveness. A 1978 report requested by President Carter to define how government could stimulate technological innovation emphasized indirect measures rather than direct government involvement. In particular, it recommended reinforced patent legislation and the repeal of antitrust restrictions in order to allow cooperative research between rival firms in the same industry.[37]

The proponents of the business ethic, who came back to power with the election of Ronald Reagan, in 1980, adopted most of these measures as their own. They called for technical innovation and a renewal of the entrepreneurial spirit, but they preferred to rely on private initiative rather than on the government. The new administration was calling for a renaissance of the business ethic. Its aversion to government involvement in the economy led it to propose major cuts in civilian R&D programs begun by the preceding administration. For example, the budget for energy research was halved between 1980 and 1985.[38]

Some agencies, such as the National Bureau of Standards and the National Institutes of Health, were spared similar treatment only because of their numerous defenders in Congress. On the other hand, military R&D budgets were greatly increased. As East-West tensions worsened, Ronald Reagan had no problem convincing Congress that a stronger military posture should be the nation's first priority. With more than 70 percent of available resources devoted to defense, federal R&D budget priorities once again resembled those of the early 1960s.[39]

In spite of these shifts in priority, there are a number of constants underlying American technology policies.

The first is the existence of a large consensus for government involvement in developing the technologies needed for national security. This consensus, and the power of the military lobby, explain why the military R&D budget, which has sometimes reached 80 percent of total federal outlays for R&D, has never fallen below 50 percent, in spite of strong swings in government R&D priorities. This does not mean that there is always agreement on objectives and programs. The controversy around SDI and the size of recent Pentagon budgets illustrate this point.

The second constant results from the dogma of government neutrality in the R&D sector. While everyone agrees that it is the government's role to finance basic research, the results of which are a "collective good," they also agree that it is not the government's role to become involved with commercial technological development, which is a matter for private firms. Within the political class there is general agreement that the government should not become involved wherever research costs can be recovered by sales of products resulting from that research.[40] This criterion does not cover everything, however. The principle is that it is a company's responsibility to decide on and to finance commercially viable technologies, but the government may step in when the expected profits are difficult for a single firm to appropriate, or when they can be earned only over a very long period.[41] These are the reasons that were given through the 1960s and the 1970s to justify the launching of certain civilian technological projects. The supersonic-aircraft project launched by Kennedy, the Clinch River breeder reactor project approved by Nixon, and the synthetic-fuels program initiated by Carter are notable examples of commercially oriented technological programs instigated by the government. Each of them stirred up a political storm. None of them was completed. Whatever the reasons for their abandonment, they left behind a feeling of failure and a mistrust for government initiatives in the civilian technology sector.[42]

These two constants—widespread congressional support for government involvement in developing the technologies needed for national security, and mistrust of public involvement in commercial technologies—explain the paradox and the seeming hypocrisy of the US government's technological policies. The government will never miss an occasion to restate the principles of a free-market economy and the need for a hands-off policy while giving the Pentagon abundant resources that will indirectly serve to shape large portions of the country's high-technology sector. Seen from Europe or Japan, American technological policy seems to be made by the Pentagon, which in effect plays the role of an American version of MITI (Japan powerful Ministry of International Trade and Industry). It is true that the Departement of Defense seems to be the only government body with the resources to meet the strong foreign technological challenge. It is also true that a progressively broader interpretation of national security has caused the Pentagon to show concern for the future of some strategically important industries that have been threatened by foreign competition. But althrough certain Pentagon programs do in fact have an impact on civilian industries, the Pentagon's image as a Japanese-style tool for technological and industrial policy is far from true, as we shall see. This notion was nevertheless implicit in many official speeches alluding to the prospect of beneficial civilian fallout from the Strategic Defense Initiative.

The euphoria that surrounded the launching of a number of military technology programs during the early years of the Reagan administration contributed to perpetuating the myth of American technological superiority. A return to reality would be painful to face.

The Mirage

6

John Linvill is a scholar unlike most others. In Stanford University's department of electrical engineering and computer science, where he teaches, it is common to encounter professors who spend part of their time working for industry; Linvill, however, is a true entrepreneur. In 1971, with three colleagues, he founded Telesensory Systems, which manufactures Optacon, a reading device for the blind that was developed in his laboratory at Stanford. Today, thanks to this invention, thousands of blind persons are able to read electronically, without Braille translation.[1] Linvill is also the creator of the Center for Integrated Systems (CIS), a model for the new cooperative research centers bringing together researchers from industry and the universities. Most of the major firms in the American electronics industry have become involved in this endeavor, including IBM, Digital Equipment, Texas Instruments, Motorola, and TRW. Twenty of these companies paid $750,000 apiece to finance the center for its first three years. Since 1986 it has been housed in an ultramodern building whose $15 million cost was shared by the university, the federal government and the industrial subscribers. The purpose of CIS is the design and manufacture of integrated circuits and the development of applications. CIS represents a unique research tool for studying the entire range of technologies used in the microelectronics industry.[2] Within a few years it had become a model for industry-university cooperation, which is part of the tradition of Stanford University.

The American University: A Unique Concentration of Intellectual Resources and Innovation

It was from the Stanford campus, in the late 1930s, that the first of the initiatives that would later give birth to Silicon Valley came. Frederick Terman, a professor of electrical engineering, was encouraging his students to create their own companies. When Terman, after a stint at Harvard, returned to Stanford ten years later as dean of the engineering department, he started the first research center associated with the university. Among the Stanford graduates who established businesses were Russel Varian, the inventor of the cathode-ray tube, and William Shockley, the inventor of the transistor.[3]

Since that heroic time, creating high-technology firms has become a tradition for Stanford graduates. More generally speaking, the university has managed over the years to weave special ties to the industrial environment, and CIS is only the most recent example. Data processing, electronics, advanced materials, biomedical science, and biotechnology are the principal areas in which these ties have been developed.

The status of a professor at a university like Stanford is enough to encourage such relations. A bright PhD, admitted to the staff after intense competition, will for the first seven or eight years hold only the precarious job of Assistant Professor. In order to become tenured after this long trial period, he must stand out by the quality of his research. But in order to carry out research work he will need, more often than not, to find the financing himself. The amount of public or private money he is able to attract is often viewed as an index of his success, and is used by new students entering the PhD program as a criterion for selecting a thesis adviser. Another aspect of the young professor's situation that incites him to develop outside contacts is the fact that he is paid by the university only nine months a year. He has no choice but to find other sources of income during the summer—for example, working as a consultant to industry.

Such conditions are not limited to Stanford. They are to be found in most technology-oriented universities that have sought to develop their relationships with industry. In this category are Stanford, Berkeley, UCLA, Caltech, the University of Texas, Georgia

Tech, MIT, the Rensselaer Polytechnic Institute, Carnegie-Mellon, Cornell, Northwestern, the University of Illinois at Urbana, and Purdue, to cite only a few. Often possessing the best teams in the world in certain technical specialties, and able to attract the best professors and students, these universities, by virtue of their quality and their number, provide a resource to be found in no other country.

Although they are a minority among the approximately 3,250 institutions of higher education in the United States,[4] the technological universities are the spearhead of the American innovation system. They cast no shadow, however, on other well-known universities, many with more fundamental research teams that belong to the worldwide scientific elite. Whereas many European countries have significant public research sectors, in the United States it is within the university laboratories that most fundamental research is carried out. This explains the impressive funding devoted to research by American universities—$9.5 billion in 1986, of which 63 percent was financed by the federal government, 5 percent by industry, and 32 percent by the universities and miscellaneous sources.[5] Able, thanks to these resources, to attract the best researchers in their specialties, American universities have become the world's greatest breeding ground for research. More than 150 Nobel Prizes have been earned there since the awards were created in 1901. Harvard has received almost thirty, Berkeley fifteen, and Stanford twelve. The award established in 1982 by the National Academy of Sciences for the best universities for research has been won by Harvard, Berkeley, and Stanford.[6] MIT is still one of the best schools in the engineering sciences; Harvard excels in fundamental disciplines as varied as physics, mathematics, medicine, history, and philosophy.

There are 13.5 million students in American universities, of whom approximately 40,000, or 3 percent, are from abroad. About 990,000 diplomas are awarded each year, including 77,000 in the engineering sciences. From the standpoint of size and quality, this system has no equal in the West.[7]

Basic research performed in the universities nourishes overall technological development downstream. Its role has grown over the years with the expansion of the scientific and technological base needed for industrial development. Microelectronics, biotechnology, composite materials, opto-electronics, and computer networks are among the generic technologies that have emerged over the last twenty years at the crossroads between the engineering sciences and the great traditional scientific disciplines such as physics, chemistry, and biology. Cross-fertilization characterizes these new domains for research. For example, genetic engineering requires not only biologists but mathematicians and computer specialists as well.

The results of such generic research are significant for industries in both technical cultures. For companies involved in competitive markets, the university laboratories represent an exceptional reservoir of ideas, know-how, and scientific and technical competence. They provide a unique means of access to innovation and technology for firms that do not have their own R&D capability. This partly explains the rapid development of collective research centers in the United States over the past ten years, such as the Center for Integrated Systems.

For the Department of Defense and for the industries in the sheltered culture, access to university brainpower is no less essential. Within university laboratories are specialists working at the leading edge of knowledge in their discipline. The Pentagon's continual effort to strengthen America's technological leadership needs their collaboration.

However, universities and military research have not always gone well together. Growing hostility to research financed by the Department of Defense emerged on campuses in the late 1960s. Resulting from the Vietnam war, this opposition caused numerous universities to turn down Defense Department research contracts, particularly for classified projects that could not lead to publication. This posed an enormous problem for MIT, where much military research had been done in the 1950s—particularly at the Charles Stark Draper Laboratory.[8] This laboratory had earned distinction through its work on inertial guidance systems. It had been given the assignment of developing the guidance systems for the Polaris and Poseidon missiles. Later, the Draper Lab was asked by NASA to construct the navigation system for the Apollo lunar mission. As one of the most prestigious military research centers, the Draper Laboratory is a prime example of the sheltered culture's outstanding capabilities. In 1969 the Draper Lab had a budget of $52 million and employed 2,400 persons.[9] Its direct participation in the development programs for new weapons made it a prime target for student demonstrations. Although other MIT laboratories, such as the Lincoln Lab, were also receiving significant military financing, only the Draper Lab had assumed such operational responsibilities. An elegant solution was found in 1970 for the dilemma posed for MIT's administration by the growing opposition to military research and by MIT's need for the Draper Lab's scientific and human resources[10]: the Draper Lab was separated administratively from the rest of the institute. Thus, without cutting its ties to the Pentagon, MIT quickly reduced its share of military contracts, keeping only fundamental research contracts. The Draper Lab, located off campus, continued to be involved in the development of new weapon systems.

By the early 1980s, memories of the Vietnam period had faded, and renewed tension between the East and the West gave new legitimacy to military research in the public eye and on the campuses. Furthermore, since the mid-1970s the Department of Defense had been increasing its financial contributions to basic research. In 1985 this financing represented 10 percent of university research. The SDI program gave strong impetus to this process. Pentagon money began to flood certain scientific sectors, such as data processing. Soon there was not a single artificial intelligence research laboratory that was not benefiting from SDI Office financing. In 1986 military financing represented as large a part of the American university system's resources as funding provided by the National Science Foundation, one of the principal financial sources for fundamental research in the United States.[11] However, more than half of the money was going to two institutions that had developed particularly close ties to the Pentagon: MIT and Johns Hopkins.

The Laboratories Innovate; Industry Stalls

Alberto Sangiovanni Vincentelli is vice-chairman of the department of electrical engineering and computer science at Berkeley, where he has been teaching since 1976. Well known in his profession for his exceptional mathematical talent, he devotes all his available time to research in the design of integrated circuit. His specialty is the computer-assisted design of very-large-scale-integrated circuits. His laboratory produced SPICE, a widely used software tool for simulating and testing circuits.

Alberto Vincentelli is always happy to discuss the latest developments in a field that in many respects bears the mark of work done in his laboratory. "Thanks to silicon compilers," he says, "today we need only a few hours to design and produce a circuit whose development previously required several weeks of engineering time. Anybody with a computer terminal comparable to those my students use can now design, simulate, and test an application-specific integrated circuit . . . that fills a user's particular requirements. This will mean a revolution in the industry, by making it possible to produce custom-made circuits at no greater cost than for mass-produced standard circuits."[12]

Born near Milan, Alberto Sangiovanni Vincentelli is typical of the brilliant European researcher seduced by the American way of life—or, more specifically, by the exceptional working conditions to be found at Berkeley: practically unlimited research funds, and the opportunity to experiment with the most advanced computer hardware available (generally provided free of charge by the manufacturers). More than anything else, he is attracted by the excellence of the students. Only one out of ten applicants is admitted to the PhD program. These are graduates of the world's most renowned schools and universities. Nationality is not a criterion for selection. "It's the only way," says Professor Vincentelli, "to select the best."[13]

It is the spring of 1986. A distance of less than 40 miles separates Berkeley from Santa Clara via the Camino Real, which runs the length of Silicon Valley. Semiconductor firms have begun to appear in San Carlos, alongside the highway and on cross-streets, with signs reading Harris Components, Ask Computer Systems, Varian Associates, and Digital Equipment. This section of the Camino Real also contains, over the span of a few miles, the highest density of computer merchants on earth. But what catches the eye even more is the number of signs advertising "Premises for Sale" and "For Rent." A study made by Grubb & Ellis, a San Francisco real-estate firm, confirms the impression: with 34 percent of its industrial facilities vacant, the area holds the record for the United States.[14]

This contrast between a campus alive with the fever of discovery and a phantom industry seems to summarize the state of American high technology, or at least as much of it as can be seen in the Bay Area at the beginning of 1986.

For the past several months this surprising image had been reflected in the financial fortunes of high-technology firms, particularly in the electronics and computer sectors. In the third quarter of 1985, Intel announced an operating loss of $23 million. National Semiconductor showed a deficit of $54 million, Commodore $80 million, and Control Data $256 million. In June 1984, National Semiconductor laid off 1,300 employees. Early in 1985, Texas Instruments announced its intention to lay off 2,200. On July 24, 1985, the *Wall Street Journal* revealed that over the first six months of the year Silicon Valley had lost 3,600 jobs.

The stock market did not wait for summer to be overcome with doubt. Wall Street's 1983 enthusiasm for high technology now appeared to have been only a flash fire. Especially in electronics and computers, large numbers of investors discovered that they had lacked discernment in backing certain companies. In 1982 any new entrepreneur in quest of capital had only to label his endeavor "high-tech"—a magic word synonymous with fast growth and an exceptional rate of profit on capital invested.

"Too much money chasing too few projects" is the way one venture capitalist described his profession late in 1984. That was the year of the rebound. High-tech shares began to be sold at discounts. Since the end of 1983 an inverse movement to the one seen beginning in the spring of 1982 had been developing on Wall Street. The Hambrecht & Quist index of high technology firms fell from 280 in the early summer of 1983 to 190 in the spring of 1985.[15] High technology was no longer the golden-egg-laying goose that so many investors had hoped for.

This pessimism in the world of business and finance seemed to find further justification in the changes occurring in the overall indicators for American industry, particularly in those for the high-tech sectors. In fact they were reflecting an alarming diminution in the competitive posture of American companies. The productivity of American industrial labor was still the highest in the world—as measured by gross national product per wage earner, it amounted in 1985 to $39,000 per year, versus $22,000 in Japanese industry.

But the lead was narrowing dangerously. The United States had experienced one of the lowest labor-productivity growth rates over the past twenty years. The decrease was illustrated by the change in the US trade balance for manufactured goods: positive until 1981, the balance has been deteriorating ever since, leading in 1986 to a deficit of $138 billion.[16]

The unchallenged supplier to the world in the 1960s, the United States has seen its position as a manufacturer of industrial products deteriorate during the 1970s and the 1980s. During this period the penetration of foreign products into the American market spread to an ever-larger number of sectors: clothing and textiles, consumer goods, electrical equipment, steel, automobiles. Several of the major branches of the industrial-equipment industry, which were still generating a foreign-trade surplus in 1980, were showing a deficit by 1981. In 1985, 50 percent of all textile machines were imported, and 43 percent of machine tools.

This loss of competitiveness in the traditional industrial sectors was nevertheless seen by some as a natural phenomenon illustrating the United States adaptation to the international division of labor. The comparative advantage of the United States in the case of products with high added value would enable the country to compensate for this loss through increased exports and access to new markets in the high-tech industries and services—a natural evolution, for the United States, toward a postindustrial economy. At first glance, the structure of American foreign trade lends weight to this hypothesis. Since the mid-1960s the proportion of high-technology products in American industrial exports has gone from slightly more than 25 percent to nearly 50 percent. The surplus in foreign exchange of these products grew threefold between 1972 and 1980. But a reverse trend has appeared since then (figure 3): as the trade deficit in the traditional industries grew, the surplus in the high-technology sectors narrowed and became a deficit in 1986. A more detailed analysis of the different branches of industry shows that, on the average, the performance of the high-tech sectors of American industry in comparison to their foreign competitors has not been as good as for industry as a whole in recent years. For example, imports of high-technology products grew faster than the average for other manufactured products, increasing by 165 percent between 1980 and 1986, while exports in the same sectors increased

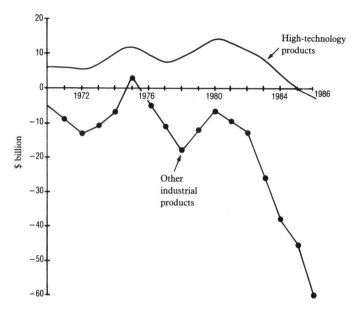

Figure 3   US balance of trade for traditional and high-technology industries (in constant 1972 dollars). Source: US Department of Commerce.

by only 29 percent.[17] For the first time, the value of American high-tech imports was outpacing what the industry was able to export.

The most valuable asset of American industry, "high tech," suddenly seemed unable to fulfill the promise—announced by the president of the United States himself—that it would bring back American industrial and economic prosperity. The myth of high technology "pulling along" the rest of the American economy seemed suddenly to have evaporated.

Stategic Defense: From Dream to Reality

Entrepreneurs and investors were not alone in the need to make a painful and costly adjustment of their hopes for high technology. A similar adjustment was soon to occur with regard to Ronald Reagan's cherished technological project, the Strategic Defense Initiative.

Reagan's speech of March 1983 had opened a new perspective in a debate that had for thirty years been the subject of a bitter controversy among specialists: the debate on the balance of nuclear forces and arms control. Domestically, the time seemed particularly appropriate for a change of perspective. Since the end of the 1970s American public opinion had become increasingly less confident about nuclear deterrence.[18] The pause in the arms race brought about by the détente of the 1970s had worked primarily to the advantage of the Soviets. New sources of tension had recently appeared between the Eastern and Western blocs, from Afghanistan to Central America, increasing the danger of a nuclear holocaust. In this context, developing defensive weapons instead of pursuing the endless race to improve offensive ones was an attractive idea. Even though in the early 1970s all experts agreed it was not feasible, the creation of an anti-missile system could be contemplated in the 1980s. The technology available during the 1970s allowed for intercepting enemy missiles only during the final phase of their trajectory, as they were approaching American territory. It was partly to offset this threat, and to counter the rudimentary missile-defense system the Russians had installed around Moscow, that the United States developed multiple-warhead missiles.[19] The technology of the 1980s made it possible to consider intercepting Soviet missiles at a great distance, while they were still in their ascent phase—that is, when the number of objects to be destroyed was still relatively small. Two scientific and technical breakthroughs had led to this radically new approach to missile defense: significant advancements in sensors, microelectronics, data processing, and telecommunications had made the identification and the tracking of missiles easier, and progress at the level of physical principles might make it possible to develop directed-energy weapons utilizing lasers or particle beams. These technical advances had played a determining role in the decision to launch the Strategic Defense Initiative.

However, in the defense industry, as in most industries that are driven by technological progress, an abyss can stand between the discovery of a physical principle or a new technical possibility and its operational implementation. There are many examples that illustrate how unrealistic some technological objectives can be. The Dyna Soar project is one of them.

In 1958, Boeing and Martin began a study for the Air Force of the feasability of building a new kind of aircraft—the Dyna Soar—which would be carried into orbit by a Titan rocket and would then glide back to earth and land like a conventional airplane. As of 1962, the Department of Defense was devoting $100 million to developing this precursor of the space shuttle. A budget of $921 million was forecast through 1969.[20] This project was abandoned in 1963 by the Kennedy administration for budgetary reasons. We might wonder what would have happened if the money had been made available to the Air Force. The goal was to make Dyna Soar operational in 1966—some fifteen years before the shuttle—although at the time none of the materials needed to absorb thermal shock from the return through the atmosphere existed.

On the other side of the coin, certain specialists feel that the speed with which new technical opportunities are exploited in the military field is, more than anything else, a question of means and political will. The classic example illustrating this view is the development of the atomic bomb. Only 3½ years passed between the moment President Roosevelt gave a green light to the Manhattan Project and the first explosion at Alamogordo.[21] At the Department of Defense, proponents of SDI often mentioned the Manhattan Project as a precedent. The SDI Office assembled an impressive list of concrete proposals for implementing the president's vision. The basic work was done by James Fletcher, a former NASA Administrator.[22] Fletcher's report confirmed the possibility of developing directed-energy weapons, with a vast R&D effort, and he proposed several research approaches.

Important progress had been made on x-ray lasers. Free-electron lasers and chemical lasers seemed especially promising. In the latter case, the laser effect would result from a chemical reaction producing very large amounts of energy, such as a hydrogen-fluorine reaction. In principle, developing such weapons would require nothing more than the extrapolation of existing lasers to very high energy levels.[23] Radiation in the infrared band, unlike x rays, would be only weakly absorbed by the atmosphere. However, the launching into orbit of complex equipment requiring an energy source of several hundred megawatts and weighing at least 100 tons remained problematic. A free-electron laser based on the ground might be an alternative. In this device, the laser beam would be produced by slowing down of electrons that had been moving at high speed in a particle accelerator. The beam would be aimed at its target using a combination of mirrors placed in low geostationary orbits.[24] Intense beams of uncharged particles could be another possibility for directed-energy weapons, but this option would require the orbiting of heavy and complex ion or heavy-electron generators.[25] Kinetic weapons could complete the SDI arsenal. They would use projectiles weighing approximately 1 kilogram, accelerated by an electromagnetic rail gun to speeds of 20 to 100 kilometers per second. Although of shorter range than the laser and particle-beam weapons, kinetic weapons would have the advantage of becoming operational sooner.

Another theme for SDI research was the study of command and control systems. An intense research effort would be needed to develop computers and software able to process the information coming from missile-identification sensors and to aim and fire the various types of weapons in the defense system.

As the strategic, scientific, and technical implications of SDI began to be discussed in the media, the program aroused skepticism among some American scientists and opposition among others. Some denounced "Star Wars" as a dangerous militarization of space and a violation of the commitments made under the ABM treaty of 1972; others doubted the feasibility of the weapons under consideration and (especially) the president's announced schedule for producing them. The program was opposed by such renowned scientists as Hans Bethe, Richard Garwin, Victor Weisskopf, and Henry Kendall, and by the Union of Concerned Scientists, an organization that some ten years earlier had crusaded against nuclear power. Petitions began to circulate in the universities.[26] There was increasing public debate, some of it organized by the SDI Office in an effort to counter the growing influence of its critics.

In October 1985 a debate was held at the Cosmos Club, a Washington organization to which many scientists belong. General James A. Abrahamson, director of the SDI Office, was the guest of honor. During his talk he sketched a comparison between President Reagan's vision of a "space shield" and the vision John Kennedy supposedly had when he decided in 1962 to launch the Apollo program. With his usual enthusiasm, the general concluded by saying: "Our success in going to the moon can be repeated today with the SDI project." At that point he was interrupted by a member of the audience who stood up and said: "There is a fundamental difference, General, that seems to have escaped your attention. Unlike Star Wars, the reasoning that led to Apollo was not in contradiction with the elementary laws of physics." This undoubtedly excessive criticism foreshadowed the bitterness of the ensuing debate between proponents and critics of SDI. Were the laws of physics being violated? Probably not, but that is not to say that the project's underlying vision was realistic, or that the technological developments it implied were feasible.

A study group created by the American Physical Society provided the most complete, best-documented, and least controversial analysis of the state of the art and the feasibility of the technologies being considered for SDI. The group made its conclusions public in a report published in the spring of 1987.[27] They were not kind to the program: "Although substantial progress has been made in many technologies of DEW (directed energy weapons) over the last two decades, the Study Group finds significant gaps in the scientific and engineering understanding of many issues associated with the development of these technologies. . . . crucial elements required for a DEW system need improvements of several orders of magnitude. . . . At present, there is insufficient information to decide whether the required extrapolations can or cannot be achieved. . . . We estimate that even in the best of circumstances, a decade or more of intensive research would be required to provide the technical knowledge needed for an informed decision about the potential effectiveness and survivability of directed energy weapons systems."

According to the APS group's report, the power of deuterium fluoride lasers would need to be increased 100 times, and that of excimer lasers 10,000 times. The intensity of the ion-beam accelerators needed for the development of particle-beam weapons would have to be increased by a factor of 100. "Furthermore," the report continued, "an evaluation of the possible use of free-electron and x-ray lasers as antiballistic weapons depends on previously validating numerous physical concepts."[28]

Another extremely complex problem was that of developing a computer system able, in real time, to process, coordinate, and integrate millions of pieces of data per second, coming from thousands of sensors, radar stations, and other installations used to identify and track enemy missiles. This information would need to be available to hundreds of weapons, on the ground and in orbit, requiring extremely precise aiming and firing.

David Parnas, a professor of computer science and a consultant to the Office of Naval Research, resigned from the SDI committee in September 1985, contending that the development of software able to manage and coordinate the SDI system was impossible.[29] A short time later, thirty computer specialists—many of them renowned for their scientific work—stated that writing the software needed for SDI would be impossible before the end of the century.[30]

Among the many requirements of the SDI project was the need to build and orbit nuclear or conventional energy sources providing dozens or even hundreds of megawatts for each directed-energy weapon in orbit. In all, the program would require assembling in space approximately 100 space stations of impressive size. This would be a titanic task, requiring 500 to 1,000 launches of the present-day shuttle.[31]

An even more fundamental issue soon forced the Department of Defense to make a major reassessment of the president's SDI vision. This had to do with efficiency of the protective shield. However optimistic one might have been, it soon became clear that a shield that would be totally impenetrable to Soviet missiles was a utopian dream. Critics stressed that if even a small percentage of the missiles launched against the United States penetrated the defense system, they would cause considerable damage. In this case, what was the value of a necessarily imperfect defense system? It would be, SDI proponents said, "an essential tool for enhancing the credibility of the US deterrent force, at a time when the East-West balance of power was threatened. A shield able to protect ICBM silos would itself help deter nuclear attack."[32] Under these conditions, SDI was at best only a bargaining chip, replied its detractors—an opinion apparently shared by a good number of experts. In any case, this limited value would seem to be the main reason why congressional cutbacks in the president's budget requests for this program—10 percent in 1985, 17 percent in 1986—were not even larger.[33]

"Some of those who voted for Star Wars financing are in favor of continued research in the hope that it will force the Soviet Union to make concessions in arms control," said one expert, adding: "The United States missile-defense program has lost much of its glamour, with members of Congress doubting the president's assertion that the program would protect the American population against missile attack. "[34]

We are indeed far from President Reagan's vision of SDI, and its underlying grand plan to rid the planet of nuclear weapons once and for all. However, missile defense—a taboo subject until 1983— has become a respectable subject for research, and an element that deterrence experts and strategic analysts now must integrate into their long-term vision of East-West relations.

When Technology Is a Mirage

The meeting to review the final preparations for *Challenger* launch 51L was scheduled for January 15, 1986. This mission had two objectives: to orbit TDRS, a telecommunications relay satellite that would make it possible for NASA to do without ground-based tracking stations, and to deploy a Spartan satellite for observing Halley's Comet. For the public, however, the star of the day was Christa McAuliffe, who had been chosen from among several dozen candidates to be the first schoolteacher to go into space. For NASA the choice of a teacher had symbolic value: it would symbolize the general public's access to space flight, and it would capture the attention of the young.

The idea of including a teacher had come from James Beggs, NASA's Administrator. He had submitted it to the White House for approval in 1984,[35] and it had been accepted immediately. With flight 51L, the shuttle's twenty-fifth launch, Beggs intended to show that the program had come of age and that NASA had made space flight routine.

Ironically, Beggs was no longer NASA's Administrator at the time of this historic launch. He had been forced to resign two months earlier because of suspicions linking him to a payoff scandal involving the Pentagon and one of its major contractors, General Dynamics. He had been president of that company before joining NASA in 1981.

Scheduled for January 22, the launch was delayed to the 26th and then the 28th. Breaking with tradition, William Graham, the acting NASA Administrator since Begg's departure, would not be attending the launch because of meetings in Washington. As the last minutes in the countdown went by, Graham was on Capitol Hill.[36] At 11:38 A.M. the on-board computer ignited the shuttle's main engine. A few seconds later the solid-fuel booster rockets ignited. *Challenger* rose slowly over the launch pad. Seventy-three seconds later, the nightmare NASA had feared for twenty years happened. *Challenger* exploded over Cape Kennedy.

America was stunned. Probably no event since John Kennedy's assassination had aroused such public emotion. Aside from the human tragedy, it was to some extent the country's dream that had gone up in smoke. For 25 years the space adventure had been an integral part of American culture. Even though public interest in the conquest of space fluctuated during this period, Americans saw the astronauts as descendants of the pioneers, and space as the new frontier. NASA had cleverly played on this image. At the Air and Space Museum in Washington, a film glorifying the shuttle and the astronauts reminded viewers that "the dream is still alive." Now the dream had vanished, and during the weeks following the accident, as the Rogers Commission investigated the causes of the catastrophe, the public began to discover the reality of the space program.[37]

The prestigious NASA of the Apollo era had become a bureaucratic organization subject to power struggles and multiple rivalries. Authority and decision-making power had become dispersed, with no clear definition of responsibilities. The organization's relationships with suppliers had prevented it from properly supervising quality control. Safety had been sacrificed to budget savings.

Such were the severe conclusions of the Rogers Commission, which devoted the greater part of its report to negligence and inadequacies in NASA's technical decisions with respect to the solid-fuel boosters, which were identified as the cause of the accident.[38]

The public also discovered the errors of judgment made by the government, which had bent to NASA's demands. No, space exploration was not an undertaking like any other. It involved human risks that would not soon disappear. To have taken a civilian aboard showed a serious lapse in judgment. Making space available to the man in the street was utopian, at least for the remainder of this century. Another mistake was to have given the shuttle a monopoly on NASA launches. This mistake was all the more serious in that a major hypothesis on which the shuttle program was based had now been proved false. The shuttle would never be the all-purpose, inexpensive means of reaching orbit that it had been proclaimed in 1972. The absurd situation in which the American space program found itself—unable to dissociate manned flights from launches of commercial, observation, or telecommunications satellites—was now clear to all. Fortunately, the Department of Defense had not fully accepted NASA's monopoly. The Air Force had continued to put some of its own satellites into orbit using the old Atlas and Delta boosters (from General Dynamics and McDonnell Douglas, respectively). Having kept this launch capability operational suddenly became crucial, as the shuttle would be grounded for at least two years.

But the string of bad luck seemed to be continuing for the American space program. On May 3, 1986, a Delta booster being used to orbit a GEOS satellite for the Pentagon exploded shortly after liftoff. A few months later, on March 26, 1987, the same thing happened to an Atlas-Centaur that was supposed to put an FLT-SATCOM satellite into orbit.[39] A year after the *Challenger* accident, the nation that had first put a man on the moon was without any launch capability, not even for orbiting a modest telecommunications satellite.

Three and a half years after President Reagan tried to communicate to the American people his vision of space technology, events had dramatically highlighted the gap between the objectives underlying that vision and the reality of the American space program. As SDI research continued, the contrast between the real and the imaginary became as clear as the light of day. The civilian and military space adventure seemed to have regressed to the sort of nineteenth-century technological perspective so well illustrated by Jules Verne. His was an enthusiastic and prophetic vision, but it was ahead of the scientific reality of his time by a century.

President Reagan's view of technology seemed similar. A mixture of awe and optimism, apparently divorced from reality,[40] it gave various interest groups besieging the White House an opportunity to obtain approval for the most futuristic projects. Never in the history of the United States have so many major technological projects been launched in such a short period of time. In addition to SDI, they include the orbiting space station, the NASP hypersonic airplane, the stealth bomber, and the Superconducting Super-collider. These large technological projects have two things in common. The first is their cost, uncertain but extremely high, each of the order of several billion dollars.[41] The second is that most of them were launched in haste (the SSC is an exception), with little preliminary consultation, and with none of the bodies that might have been able to evaluate them called upon early enough. Each of these projects is a technological and financial adventure.[42]

On August 11, 1986, by 239 votes to 176, the House of Representatives decided to reduce the 1987 SDI budget from the $5.3 million requested by the White House to $3.1 million—a 41 percent reduction. Representative Roy Dyson summarized the House's state of mind when he said, "I don't know of any other major defense program put forward without knowing what it was supposed to do, how it would work, or what it would cost."[43]

A comparable distance between expectations and realities had recently appeared throughout the American high-technology sector. It seemed to affect both of the technological cultures. At the end of 1986 Silicon Valley industrialists and Pentagon bureaucrats were sharing the same concern. It was as though they had both discovered they were victims of the mirage of technology.

For industries in the exposed culture, the mirage expresses itself economically and financially in terms of bankruptcies, losses of income, and lost market shares. Its consequences are particularly dramatic for industries facing strong international competition. In emphasizing high technology, have American companies not dropped the prey for the shadow? The mirage of technology is a threat that now hangs over the prosperity of America and its capacity to hold first rank among industrialized nations.

For industries in the sheltered culture, with few exceptions, the mirage is not synonymous with significant financial losses. Government financing for large military and civilian technological programs will not disappear overnight, regardless of the problems. But the situation of these industries is far from comfortable. The mirage will probably lead to questioning and reorientation of the government's objectives and priorities. This is the worst thing that can happen in a planned universe in which the measure of success is precisely to have reached the assigned objective, to have attained the projected performance, to have achieved the technological prowess acquired secretly through long months in laboratories. By having set unattainable goals for the industries of the sheltered culture, the White House, the Department of Defense, and NASA have cast suspicion and doubt on a system whose main source of pride has always been its ability to overcome the technological challenges set by political leaders. This change suddenly threatens the sheltered culture with the loss of what it needs most in order to accomplish its mission: the support of public opinion. A crisis of confidence and widespread disarray in the sheltered culture are the most serious consequences of the mirage of technology.

A Technological Challenge?

American companies in the exposed culture were soon able to identify who was responsible for the mirage of technology. Japan was clearly the villain. Sony, Hitachi, Matsushita, and Toshiba had been visible in American high-technology markets for some years, and had penetrated them with awesome efficiency. Japanese products were nothing new in the United States, however. Japanese cars had begun to appear in the early 1970s, and the names Honda, Toyota, and Datsun were as familiar to American consumers as Ford and General Motors. The only thing limiting this invasion had been import quotas. Japanese consumer-electronics products, such as television sets and hi-fi components, were also becoming more and more prevalent.

This development did not cause much concern, however. Japan in the 1970s was one of the countries with a recognized comparative advantage in international competition: cheap labor. It was no more troubling to see part of the consumer-electronics industry leave the United States for Japan than it was to see the progressive shutdown of American textile and steel mills to the benefit of those in the Third World. These were natural changes in a free-market economy where America's comparative advantage lay in the development of its specialization in high technology and services.

Some concern did arise in 1978 in Silicon Valley industrial circles, however, when the first Japanese chips began to be sold in significant quantities in the United States. At that time, American firms fully controlled the US semiconductor market as well as half of Europe's. This domination applied to all categories of integrated circuits, including random-access memories (the simplest mass-produced consumer products), erasable programmable memories (EPROMs), and many kinds of logic circuits and more complex microprocessors. A technological race was under way among American RAM manufacturers. Mostek, Texas Instruments, Motorola, and Intel were running neck and neck to bring out new generations of products. In 1973, 4K RAM memories went on the market for the first time, and 16K memories were introduced by Intel and Mostek in 1976.[1] Sales of integrated circuits jumped from $500 million in 1970 to over $2 billion in 1978, and they continued to increase at an unprecedented rate. By 1979 American manufacturers could hardly meet the demand. The Japanese took advantage of this situation to introduce their products. By the end of 1979 they controlled 42 percent of the sales of 16K RAMs, the most important segment of the memory market.[2]

In March 1980 another worrisome bit of news hit the profession. During a speech in Washington, Richard Anderson, one of Hewlett-Packard's managers, made public the results of a comparative study of American and Japanese 16K RAM chips. He revealed that the best American products had a failure rate six times that of the poorest-quality Japanese products. The attendees at the meeting were dumbfounded. The Japanese were now able to produce highly sophisticated electronic components more efficiently than American companies.[3]

Events during the following years turned this concern into a phobia. American manufacturers were rushing to develop 64K RAMs. This technology was to become the basic tool of the computer and telecommunications industries. Sales prospects were bright. Japanese firms jumped into this market in 1981, and by 1984 they had grabbed 60 percent of what was now a $2.7 billion market. A few months later they were able to market the new generation of 256K memories, close on the heels of American firms. Using an aggressive price policy, they were soon controlling 92 percent of the market.[4] By the end of 1984, three Japanese firms—NEC, Hitachi, and Toshiba—were among the world's five leading semiconductor manufacturers. Out of fifteen American manufacturers in the memory market in 1976, only five remained in business.

Is Japan the World's New Leader in Technological Development?

By 1985 the Japanese breakthrough in the memory market had turned into domination. In 1986, Fujitsu was the first semiconductor producer to announce a megabit DRAM, with a million memory sites integrated on less than a square centimeter of surface. In this particular field there was no question that Japanese firms had become the world's technology leaders. Although IBM, not Fujitsu, had built the first "megachips," the American company did not sell its chips; it used them only to fill internal needs.

Japanese semiconductor manufacturers did not limit their ambitions to the market for RAM memories, the simplest mass-produced electronic components. In 1984 they began exporting huge numbers of EPROMs to the United States. The price of EPROMs on the US market dropped spectacularly in the spring of 1985. Only a government-to-government agreement limiting the prices and the quantities of Japanese semiconductors sold in the United States saved the American producers' market share.[5] The Japanese companies also began manufacturing more complex products, such as microprocessors. In 1985 Toshiba began marketing a 16-bit unit, and in late 1986 it announced the introduction of a 32-bit microprocessor, thereby becoming a competitor for Intel's and Texas Instruments' top-of-the-line semiconductor products.

These changes in the market for electronic components illustrate a more profound phenomenon affecting most of the information-technology industry: in every branch of the industry, a trade deficit soon appeared between the United States and Japan (see figure 4). Even in computers, the less-questioned domain of American supremacy, a trade deficit showed up by the end of the 1970s. It was clear, nevertheless, that imports of bottom-of-the-line products—disk readers, printers, keyboards, and other peripheral computer equipment—were chiefly responsible for this deficit. Imports of Japanese computers remained marginal.

The presence of the giant IBM, which for many years had been able to set the rhythm of technical change and the standards for the entire industry, strongly countered the the Japanese computer manufacturers' international projects, at least for a while.[6] Their limited

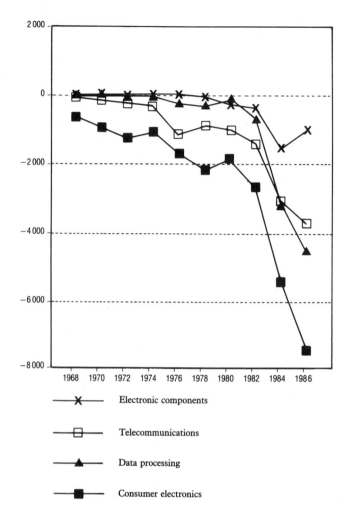

Figure 4   US-Japan trade balance (in millions of dollars). Source: CEPII-CHELEM.

penetration can also be explained by some of the Japanese companies' weaknesses, particularly in software. The Japanese eventually succeeded in producing a full range of computer systems, nonetheless. In 1984 Hitachi even announced its first supercomputer. Its arrival in a market previously monopolized by the American firms Control Data and Cray was followed closely by the arrival of similar models from NEC and Fujitsu. NEC's sale of a supercomputer to the University of Houston in 1986 was a bolt from the blue.[7]

However, what most intrigued and worried the American computer professionals was the Japanese government's creation, in 1982, of a program to do research on so-called fifth-generation machines. The announced objective was ambitious: to build "intelligent" computers able to handle complex problems as a human brain would, with a built-in learning capacity and with the ability to work in a complex information environment. The true objective, it seems, was more modest: to bring university and industry teams together to work on advanced software problems.[8]

In the US telecommunications market, dominated by AT&T, Japan's penetration was first limited to PBX and consumer-premise equipment—telephone sets, answering machines, and other equipment that could be connected to the network by individuals or companies. But in 1985, NEC began trying to sell large digital switching equipment to regional Bell operating companies. Transmission is also a strong point for Japanese telecommunications firms. They have developed remarkable capacities in optical fibers.[9] Finally, the Japanese seem to be well placed to develop advanced information technologies using opto-electronics. This field involves materials that can transmit, amplify, and switch luminous currents, as is now done with electrical currents. At stake in this research is the ability to produce ultrafast computers in which information can travel at the speed of light. Gallium arsenide is one of the more promising materials for opto-electronics, and the Japanese seem to be leading the world in this field of technological development.[10]

In 1986 the (US) National Academy of Engineering published a report of a study trip made to Japan by a group of American specialists in advanced materials for electronics and opto-electronics. The report concluded that Japan was ahead of the United States in research in seven of nine key technologies crucial to the future of this sector.[11]

Japan has also made a great deal of progress in developing advanced materials for applications other than electronics, such as composite materials. The first materials of this kind appeared in the 1960s in aerospace. A glass fiber embedded in a thermally hardened polymer matrix had already made it possible to manufacture strong, light products with complex shapes. Since that time, the resistance of these materials to mechanical stress and high temperatures has improved considerably thanks to new fibers (carbon, polymers, ceramics) and new matrices (polymer, metal, ceramic). There are many potential applications for these new materials, because of the wide variety of characteristics and properties that can be produced. While the United States has maintained first place in basic research in these new materials, American companies are now faced with competition from the Japanese, who are often ahead in certain applications—particularly in ceramics, for which one potential user is the automobile industry.[12]

In the life sciences, American's scientific and technological leadership remains unchallenged, thanks in large part to a long-term, large-scale basic research effort in agriculture and medicine financed by the government through several federal agencies. For 1987, the National Institutes of Health alone had an annual budget of more than $6 billion. The American preeminence in the life sciences is also due to an efficient system of transferring technology to industry, via a multitiude of new firms created since the beginning of the 1980s (notably in the field of biotechnology). A sizable fraction of these companies is now controlled by large agro-industrial and biomedical groups, which gives them privileged access to new technologies.[13]

Japanese scientific competence in the life sciences is found chiefly in agriculture, fermentation, and antibiotics. Aware of its shortcomings in the life sciences, Japan has undertaken a vast R&D effort. Numerous ties have been established with American universities and federal laboratories. In 1987, more than 300 Japanese researchers were working under long-term contracts with the NIH. Between 1980 and 1985, Japanese firms signed more than 200 agreements for cooperation, joint ventures, or acquisitions with small American biotechnology companies.[14]

More recently, the Japanese government's Human Frontiers project, launched with a call for international cooperation in the life sciences, has raised skepticism and concern in the United States—skepticism in the face of Japan's announcement that it was prepared to share the results of its fundamental research in a field in which it has thus far received much more than it has given to the international scientific community, and concern over the danger of seeing this cooperation lead to Japanese theft of American scientific and technical resources (in a repetition of the approach Japan used with such success twenty years earlier in the semiconductor field).[15]

The American Research System on the Defensive

The emergence of Japan as a scientific, technical, and industrial power should not have surprised Americans. Since the end of World War II, the United States has had a continuous policy of openness and exchange with Japan, particularly in science and technology. The original reasons for this were both economic and political. The effort to assist Japan to rebuild its industry stemmed from the same logic as the Marshall Plan for Europe. For American companies, the opening of new foreign markets would take up where the war effort—which had so greatly benefited the American economy—left off. But even more important, the logic of American aid to Japan served a strategic and political objective. After Yalta, Japan quickly emerged as an essential base for the American effort to contain communist influence in Asia, and as the pivot of American foreign policy in that part of the world. During the postwar period it became a priority of American foreign policy to anchor Japan firmly in the Western world and, at any cost, to keep it in the American sphere of influence.[16] Scientific and technical cooperation with Japan fell into this framework. It was also part of a more global picture. Sharing with its allies the basic knowledge needed for industrial development was a means whereby the United States could contribute to the development of the entire Western world.

Unlike Europe, Japan possessed no real scientific base at the end of the war. Thus, it benefited greatly from the American policy. Cooperation in the biomedical field began in 1945 with the creation of the joint commission responsible for studying the consequences of the bombing of Hiroshima and Nagasaki. This program was subsequently expanded and extended to cover infectious and endemic diseases in Asia. A major agreement on cancer was signed in 1974. This policy was applied to other scientific disciplines during the 1960s. In 1961 the National Science Foundation signed the first international agreement with its Japanese counterpart. Japan also became America's first partner in space research. At the end of the 1960s Japan began to benefit from technology transfers in expendable launch vehicles (Thor-Delta technology), telecommunications, and television satellites that had been granted to no other country.[17] In spite of the restrictions surrounding them (military use was prohibited, and prior authorization was required for Japan to launch a satellite for another country), these technology transfers contributed greatly to the Japanese space program.[18]

The American university system has a long tradition of relations with Japan. There are more than 20,000 Japanese students in the United States, of whom 5,500 are in California. In 1983, some 7,500 Japanese scientists were given visas to the United States. Japanese firms financed a large number of these researchers. These companies have made access to the American university system a priority for acquiring the basic scientific and technical knowledge needed for their development; that is why they spend more than $200 million annually in research contracts with American universities and on scholarships in the United States for their researchers. Numerous Japanese firms also participate in American universities' industrial liaison programs. For example, MIT has initiated active cooperation with Japan, and sixty Japanese companies participate in its industrial liaison program.

Not until the mid-1980s were the consequences of this policy perceived in the United States. The breakthroughs Japanese companies were making in American high-technology markets were correlated with the systematic Japanese policy of acquiring scientific and technical information and patents on American territory. Although Japan is far from being the only country to have such a policy, cooperation with Japan brought to light an important singularity: it is very difficult for American laboratories to obtain reciprocity in the exchanges. Japan acquires scientific information from the United States, but the Japanese institutions involved in cooperation, whether public or private, are rarely inclined to allow transfers in the opposite direction.[19] Members of the political establishment have claimed that the American research system is being subjected to intolerable pilferage. This, they observe, is because the American system is too open; foreign access to scientific and technical information should be limited.

This point of view echoes the one expressed for other reasons by the Department of Defense. Beginning in 1980, the considerable growth in the proportion of public research funds coming from the Pentagon gave new pertinence to the debate on controlling technology transfers to Eastern-bloc countries. Soon, certain sessions at professional scientific colloquia began to be closed to foreigners—particularly at meetings dealing with "sensitive technologies," such as metal-matrix composite materials or gallium arsenide. The reason given for these exclusions was concern for reducing the risk of leaks to the USSR. This practice provoked numerous protests within the scientific community.[20] The openness of the American research system is in fact one of the conditions for its effective operation and efficiency. The stiff competition between university research teams in many scientific fields is acceptable to the researchers only if their results are publishable.

Growing discrimination against foreign scientists became noticeable after 1985, although it was not always possible to tell whether the real reason for it was national security or protection of American firms against foreign competitors. William Graham, the president's Science and Technology Adviser, brought the confusion to new heights when, in July 1987, he forbade foreigners to attend a large forum in Washington on high-temperature superconductivity. The applications of this discovery, which caused great excitement throughout the international scientific community in late 1986, are essentially a long-term matter. High-temperature superconductivity was, at the time, a subject only for fundamental research. Some felt that Graham's decision was a gross misjudgment. Although such an interpretation cannot be discounted, the decision does reflect recognition of the strategic importance of certain civilian applications of research (even if still in an early stage) and the need to take care in sharing it abroad. It also illustrates the paranoia generated among some American R&D managers by Japanese competition.

This event has symbolic value: it summarizes a number of changes that have occurred over the past few years in America's exposed technical culture. This culture was able to develop and prosper only in close symbiosis with a university research system whose chief virtues were openness and free access.[21] The only difference between information that was free and available to anyone and proprietary information was the possibility of taking out a patent and exploiting it commercially. Otherwise, everyone could make use of the flurry of ideas and innovations that stood, above all, to benefit American companies. Globally, this system could function for two reasons. First, since the United States was at the most advanced frontier of world scientific knowledge and technical innovation, its dominant position was a condition for its openness and generosity toward the rest of the world. Second, American firms have always been in the best position to exploit these resources.

The Japanese threat seemed to show that the main beneficiaries of the American system of research and technical innovation were now foreigners. The Japanese challenge is not only a challenge for companies in the exposed culture; it is a challenge to the entire research system that nourishes that culture. What has been correctly perceived in Washington is not so much the objective reasons for Japan's new presence in American high-technology markets as the threat to something that is essential for the system's survival: its ability and vocation to ensure world scientific and technical leadership.

A Challenge to the Sheltered Culture?

American technological supremacy is not challenged only by Japan. For the past several years European firms have also shown up in various high-tech sectors in the United States. Among the European countries, Germany in particular has an image of solid technical competence. But European technological competition is not frightening. There is even a tendency to minimize it, except in a few particular fields, as was shown in a 1987 survey of a representative sample of American industrial and government leaders who were asked how they perceived foreign competition in high technology.

The results of this survey showed that, although Europe is often cited as a center of scientific and technical excellence, there are very few areas in which Europe's competition is seen as more dangerous for the United States than Japan's.[22]

One of the questions was "In your opinion, for each of the following scientific and technical areas, which country is the strongest US competitor?" Japan was named first in six out of ten sectors, with the following percentages of responses: 81 percent for computers, 77 percent for robots, 59 percent for telecommunications, 41 percent for new materials, 39 percent for software, and 18 percent for biotechnology. Nuclear energy, space, aeronautics, and medical research were the only sectors in which a European country was named first. For the first three, the country named was France.

The results of this survey would certainly have been very different a few years ago. The appearance of European firms in American markets is as recent as that of the Japanese in computers and electronic components. In fact, while Japanese firms were challenging American exposed-culture industries on their own terrain, some European companies suddenly raised another challenge. In attempting to compete with the sacrosanct American technological "establishment" (the large companies tied to the sheltered culture), European aerospace companies in the space and aeronautics sectors at first aroused incredulity—even irony—among American leaders in industry and government. They would soon be forced to revise their opinions.

The news broke in May 1977 that Eastern Airlines had decided to lease four Airbus A-300 aircraft for six months, beginning the following fall. The American company would be testing the European plane before deciding to make any purchases. For Airbus Industrie, the consortium manufacturing the A-300, the potential Eastern order represented an opening into the American market after a difficult period. Since the introduction of the A-300 in 1974, the world civilian-aeronautics market had been depressed. Airbus Industrie had sold practically nothing since the Air France purchase of December 1975. But 1977 was beginning under more favorable auspices. Thai Airways had ordered four planes; Air France and Air Inter each bought two planes in June, and in December SAS signed for two firm orders and ten options. On April 6, 1978, Eastern

Airlines purchased 23 aircraft, bringing total Airbus sales to 95 plus 38 options.[23] The end of 1978 saw a confirmation of the European consortium's commercial breakthrough, with overall sales of 176 planes. During the year just ended, 70 planes had been sold.

Airbus Industrie did not wait for this success to diversify its product line. Introduced in 1976, the A-310, with a smaller capacity than the A-300, went into service in 1981. Initial studies on the A-320, a new concept for a 150-seat long-range jetliner, were started in 1979. But finding a niche in an already crowded market would be a delicate process. Production did not start until in 1984, when the Franco-American joint venture between SNECMA and General Electric became able to provide Airbus Industrie with a high-performance engine derived from its CFM-56 and adapted to the A-320. The new member of the Airbus family made its maiden flight in early 1987. The number of orders and options that had already been obtained when it went into service presaged a brilliant commercial future for the A-320. At the end of 1987—a record year for sales, which reached 437—Airbus Industrie was nearing the objective it had set eight years earlier: to control 30 percent of the world market. During this period the market share for Boeing, Airbus' principal competitor, fell from 80 percent to 40 percent.[24] Meanwhile, the market had expanded. Boeing lost market share but continued to gain in sales volume. This satisfaction did not prevent the American giant from reacting strongly to the Airbus, particularly in the American market. Boeing accused Airbus Industrie of "unfair competition."[25]

Boeing rallied the federal government and the Congress to its cause. On several occasions the Secretary of Commerce asked European governments that were members of the Airbus consortium about the type of financial assistance being given to Airbus Industrie, both for developing its planes and for promoting export sales. American accusations of unfair practices became louder in 1987, when the European governments announced their decision to make a long-term loan of $3 billion to Airbus Industrie to finance the development of its next generation of airplanes, the A-330 and the A-340.[26] The Boeing-Airbus battle was thereby displaced onto the terrain of future technologies—an area in which Boeing, as a pure product of the sheltered culture, had always seen itself as an unchallenged world leader. But times, it seems, had changed. Airbus

Industrie announced the development of a new generation of aircraft while the design of the competing Boeing model, the 7J7 "propfan," was reputedly still in a preliminary stage on the drawing board. This put Boeing on the defensive, but it was still determined to continue the fight. As the *New York Times* noted, "A great deal is at stake in the battle between Airbus and the US manufacturers: gigantic investments, thousands of jobs, technological leadership and national prestige."[27]

European activity in the space sector soon attracted more American attention to European technology. Those who had only glanced at the first space developments in Europe discovered this new European competence on February 15, 1979, when Intelsat, the international telecommunications organization, announced its decision to use the European Ariane launcher to put two of its new Intelsat V satellites in orbit. Created in 1964 from an American proposal, Intelsat, an independent organization reponsible for installing telecommunications equipment and handling international telecommunications traffic, has its headquarters only a few miles from the White House and NASA. The Intelsat order was important news for Arianespace, the company promoting the European launch vehicle. Not only would Intelstat be the company's first North American customer; in addition, this would be its first truly commercial sale.[28]

Ariane placed its first Intelsat satellite in orbit in October 1983. After seven launches, five of them successful, Ariane seemed to have reached the commercial stage. Once the European launcher's reliability was seen to meet industry standards, launch orders began to come in. Within the industry, Ariane's technical advantages over its principal rival, the American space shuttle, also began to be acknowledged. Whereas the shuttle was intended for low orbit, the European launcher had been optimized to put payloads into the geostationary orbit used by all telecommunications satellites. The result was a simpler launch procedure, a point that would later become crucial when recurrent problems were discovered in the perigee engine, used to move satellites carried by the shuttle from low to geostationary orbit. In addition, while NASA was attempting to increase its launch rate, shuttle customers discovered that its schedule was virtually unpredictable. The complexity of manned space flight increased the risk of last-minute delays. As a result,

Ariane found a growing number of supporters among communi-cations-satellite professionals.[29]

Little by little the European launcher began to fill a "techno-logical niche," in accordance with the logic of market organization in the sheltered culture. It succeeded not because it was selling a service less expensively—satellite-launching tariffs, whether for Arianespace or for NASA, can in no way reflect all the true costs—but because it was filling a need with specific technical characteristics.

For the industries in the sheltered culture, the European challenge appeared to be relatively easy to overcome. Wasn't the problem like the endless arms race between the US and the USSR, in which each side strove to produce higher-performance technological sys-tems than the adversary?

For the industries in the exposed culture, the situation was radically new. Accustomed to fighting among themselves for mar-ket shares, these firms suddenly discovered, like those in the shel-tered culture, that they had a common foreign enemy. Above and beyond the new solidarity that would appear, there would be a great temptation to call for help from a previously disdained source: the government. Federal help could be solicited in several ways. The most traditional approach was used for the first time in 1985: a call for protection. When the Japanese semiconductor companies started a price war, American manufacturers accused the Japanese of unfair competition in an attempt to persuade the Reagan admin-istration to impose import quotas and price limitations on the Jap-anese government. A new approach came next: a call for public financing to offset the growing cost of technological development in certain sectors, such as memories, so that the American firm might catch up with Japanese firms in this area. The Pentagon would soon be solicited to finance a "first" for the exposed culture: the Sematech project, a joint venture between the major US inte-grated-circuit manufacturers and the government.

Thus, the Department of Defense was called upon, directly or indirectly, to contribute to meeting the challenges that had been raised for both cultures. In order to assess the relevance of the answers that were given to these requests, we now need to look at the relationship betwen civilian and military technological development in the United States.

The Role of the Pentagon: From Appearance to Reality

During the first twenty years following World War II, there is no doubt that a large number of the innovations that emerged in civilian markets originated from technologies developed through the war effort, or through military and space programs related to the Cold War. This was the case, as we have seen, for technologies as different as nuclear reactors, telecommunications satellites, air-traffic-control systems, semiconductors, integrated circuits, the first computers mainframes, commercial jets, the first composite materials, and digitally controlled machine tools, to list only the major ones.[30]

In some cases, adapting military technology to civilian needs demanded considerable R&D effort. More than twenty years of costly research and development stood between commercial pressurized water-cooled nuclear power plants and the submarine engines from which they were derived.[31] But in most cases the development was much faster, for there was considerable similarity between the military products and those intended for civilian markets. No more than ten years separated the first experimental use of jet engines in combat aircraft and the first commercial jets. It was the same for the first telecommunications satellites and air-traffic-control systems. It is important to note that these technology transfers benefited the sheltered commercial sector. Going from the military to the civilian sector is all the more easy if it implies no cultural change for the firms involved. The very notion of "fallout" reflects a process that is familiar in the sheltered culture: the process whereby an attempt is made to find some new use for an innovation or a product developed for other purposes. Furthemore, even when adapted for the civilian market, technologies stemming from the military sector are often highly technical and expensive. They give rise to products able to fill a technological niche rather than to mass consumer products.

The barrier between the two cultures is real. Firms operating within the sheltered culture, as we have seen, have great difficulty entering competitive markets. In fact, we could say that exceptional circumstances are needed for military or space technologies to give birth to products that are directly usable within the exposed culture. Such circumstances occurred several times during the 1950s and the 1960s, as we saw in connection with the history of semiconductors, integrated circuits, and the first computers. Aside from the similarity between technologies meeting civilian and military needs, another crucial element was the attitude of a certain number of firms that were the vectors for technology transfer from the military to the civilian sector. It is important to remember that these firms—Texas Instruments, Control Data, IBM—were not among the Pentagon's usual suppliers. Their integration into the sheltered culture was relatively recent when they had to make the strategic choice between remaining under the wing of the government and entering the race for civilian markets. This may be why they were able to swing to the other culture. Let us also recall that aside from these few firms that made technology transfers to the civilian sector possible, most of the government's subsidies and procurement money went to the Defense Department's usual contractors, which were unconcerned with making transfers to the civilian sector, or unable to do so.

Since the mid-1960s a growing divergence between products developed to satisfy military and civilian needs explains a great deal about the decreasing impact of military programs on the exposed culture. The skills and the know-how developed by the firms in the sheltered culture in order to be able to produce the increasingly sophisticated high-performance products demanded by the Pentagon, and the scant attention paid by these firms to the costs of manufacturing and production, stand in ever-greater contrast to the skills needed for survival in the exposed culture. Above all else, these skills involve development of the most efficient products in a competitive environment, which in most cases means reducing manufacturing costs. Over the long run, innovation in manufacturing processes weighs more heavily than product innovation. To keep up with the competition, all firms in the exposed culture have to devote increasing financial resources to R&D. For example, the amount of their own resources spent on R&D by American data-processing firms rose from $1.5 billion in 1965 to $7 billion in 1986.[32]

The increasing distance between the products of the two cultures, and the growing importance of self-financed R&D in the exposed culture, means that technological innovation in this universe has less and less to do with innovation in the sheltered culture. The image of military programs exploring the leading edge of knowledge, and thus being "in advance" of civilian technologies, seems increasingly out of keeping with reality. Drawn toward different objectives, companies in the respective cultures tend to explore different frontiers of knowledge. The result is that the very notion of "fallout" needs revising. Many experts claim that in electronics and data processing there is more fallout from civilian developments to military technologies than the other way around.[33] This opinion has been expressed by Erich Bloch, director of the National Science Foundation and a former vice-president of IBM.[34]

This is not to say that within the sheltered culture there are no further technology transfers from the military sector to the civilian. There are still important transfers of this kind in the aircraft industry, particularly in the case of engines. The American portion of the CFM-56, the engine produced cooperatively by SNECMA and General Electric, is derived directly from the F-101 engine that the American company developed for the B-1 bomber. The French portion of the new THR Propfan engine, the joint production of which was agreed to by the same partners in 1986 and which will replace the CFM-56 in 1993, makes use of technologies developed for the M-88 engine that powers the Rafale fighter. Though not all the new civilian engines are directly descended from military ones, it is extremely rare for a technology or a design method used for the first time in a civilian engine not to have been previously tested and developed for military engines.[35]

Ground-based or airborne electronics systems used in military aeronautics also generate technology transfers to the civilian sector. The introduction of automation in cockpits and in air-traffic-control facilities and the recent advances in flight simulators are examples. Electronic flight controls, for a long time used only on military aircraft, are now in use in some civilian airliners. Finally, a number of high-performance materials developed for military airframes, engines, or components or for space vehicles, are progressively being applied in commercial aeronautics, and sometimes in other civilian markets. For example, carbon-carbon composites developed for military rockets are now being used in brake systems for Formula 1 racing cars. This example shows that when it exists, civilian fallout from military or space programs usually corresponds to a technological niche that requires products with very specific performance. The fact that the high cost of such products is not necessarily a handicap is in keeping with the logic of the sheltered culture. The firm possessing this know-how may fill its niche without risk of competition, its know-how serving for a time to protect it against the entry of other firms. This does not prevent such innovations from being distributed later in other industrial sectors once appropriate adaptations have been made. Their link with the military program from which they derived then becomes increasingly slender and difficult to determine.

Comparable examples of transfers are to be found in a number of other sectors of the sheltered culture. In telecommunications, for example, transmission technologies developed to satisfy military and civilian needs have much in common. Although the technical performance and the characteristics required in the two sectors may be very different, the experience and the techniques acquired in developing military telecommunications satellites can also be used for civilian satellites. The same can be said for launchers. There is no doubt that Martin Marietta, which in 1986 decided to build commercial booster rockets, made use of experience gained in developing Titan boosters for the Air Force.[36] More than technical resemblance between products, the membership of buyers in the same technical culture and the sameness of the work of conceiving new products may explain why technology transfer from the military to the civilian sector occurs more readily within the sheltered culture than within the exposed culture.

But the influence of military programs on civilian technological development is not limited to fallout on civilian markets. By "irrigating" basic and industrial research, military research may also have a direct effect on civilian technological development. Along this line, it is interesting to examine the impact of a certain number of R&D programs launched in the early 1980s by the Defense Department's Division of Applied Research and Advanced Projects. DARPA has for many years been the spearhead of American military research, the Pentagon's "think tank" for long-term projects and for the development of the most advanced technologies. Free of short-term responsibilities, having developed a policy of scientific excellence, and able to attract the most brilliant researchers, this division has always been in a category of its own within the Department of Defense.

Several programs launched by DARPA at the beginning of the 1980s—including VHSIC (Very High Speed Integrated Circuits), a research program in advanced electronic components, and SCI (Strategic Computer Initiative), a research project focused on intelligent machines—were sometimes interpreted as American replies to programs of Japan's MITI. (VHSIC was launched in 1980 with a budget of $680 million over a period of eight years[37]; SCI was initiated in 1983 with a budget forecast of $950 million over ten years.[38]) In fact, however, each of these programs reflected a very specific Defense Department need.

The purpose of the VHSIC program was to develop the next generation of integrated circuits needed for weapon systems and equipment, particularly for missile guidance, for highly sensitive radar units, and for jam-proof communications systems. The announced objective was to make components 10 to 100 times faster, 10 to 100 times more reliable, and 5 to 10 times more compact than existing ones.[39] In contrast with earlier Department of Defense research programs, it was not only to involve the DoD's usual contractors; it was also to bring in large manufacturers of electronic components from the commercial sector—in other words, from the exposed culture. DARPA thus hoped to benefit from the know-how of large commercial firms such as IBM, Intel, and National Semiconductor. In return, these firms would be able, according to DARPA, to use the results of research financed under this program in developing their own products.[40]

Seven years later it was clear that the objective had not been met. Even when the program's results seemed to meet fairly well the forecasts that had been made with respect to satisfying strictly military needs (for example, by 1989 TRW was to be producing a high-performance "superchip" concentrating 35 million transistors on 2 square inches[41]), the research carried out for the VHSIC program quickly diverged from the concerns of the commercial firms, most of which gradually abandoned the program. Nathan Rosenberg notes: "The program reinforces the impression that the military and civilian sectors confront increasingly divergent needs. The unique performance requirements of the military, whose products must perform reliably in a highly hostile environment, has led to many costly trade-offs (in circuit design) that are of little use or relevance to manufacturers of civilian products."[42] Furthermore, the Pentagon's policy imposing strict secrecy around the research discouraged publication of the program's results. Finally, the highly specialized components developed for the military led to an emphasis on custom production techniques instead of on methods for producing large quantities at minimal cost.[43]

The original objective of the Strategic Computer Initiative was no less ambitious than that of the VHSIC program. A progress report by DARPA four years after the program's inception recalls that the goal involved nothing less than "increasing our national security and economic strength by developing a broad base of machine intelligence technology."[44] In operational terms, the objective was to introduce a new generation of battlefield robots. The program included the development of an autonomous armored vehicle, able to recognize its environment and to help its remote operator initiate retaliation to any hostility it encountered. It also included the development of expert systems that would provide real-time assistance to fighter pilots in combat. Another of the program's objectives was to create a new generation of "intelligent" weapons—that is, weapons capable of ever-finer terrain reconnaissance in their operating environments, so as to better identify the targets for which they were programmed.

These objectives led DARPA to undertake considerable R&D efforts in a large number of generic technologies and scientific fields, including parallel data processing architecture, artificial intelligence, sensors, and visual and vocal recognition systems. The breadth and the diversity of these topics contributed to the spread of the idea that the Strategic Computer Initiative would benefit a large portion of the computer industry. In addition, the involvement of specialists in man-machine interaction and advanced data-processing architecture suggested comparisons with Japan's program in fifth-generation computers.

We find in the Strategic Computer Initiative the traditional characteristics of DoD programs. Most of the resources are used to create prototypes and complex technological systems filling the DoD's specific needs. It would be premature, however, to come to any definite conclusions about a program whose impact will not really be felt before the 1990s. What can be said is that the Strategic Computer Initiative has had considerable influence on the organization of basic research in certain scientific fields. For example, the DARPA funds represented a quadrupling of the federal money available to artificial intelligence labs.

Recent evaluations of the potential fallout from the Strategic Defense Initiative have come to similar conclusions.[45] At the end of 1988, approximately $12 billion had been spent—a considerable sum of money for a research and development program. Most of these funds enabled the Pentagon's usual contractors to pursue their quest for the most advanced technologies. Whatever the fate of the program under the Bush administration, there is no doubt that some progress has already been made in new defensive weapons, even though deep disagreements persist about the extent of this progress relative to the initial goals announced by Ronald Reagan in 1983.[46] It is also clear that a certain portion of the funds spent— perhaps 10 percent, or on the order of a billion dollars—has served to "irrigate" the American research system, including the universities. It would be excessive to claim that this will have no influence on civilian technological development in coming years. Artificial intelligence, new computer architectures, opto-electronics, lasers, sensors, and advanced composite materials are among the research sectors irrigated by SDI funds. But detractors within the scientific community still question the usefulness of this irrigation.[47] They feel that it is responsible for a strong polarization in research work as a sole function of long-term military needs. This distortion would work at the expense of programs that could better serve industry's priorities. These drawbacks are such that one might wonder about the global effectiveness of the program in stimulating civilian technological development. "For purposes of discussion," notes Nathan Rosenberg, "one could pretend that SDI is mainly a federal R&D support program such as those administered by the [National Science Foundation]. One can then evaluate its usefulness to the civilian economy in the same way as one would evaluate any other program. One point emerges with great force when the problem is viewed from this perspective. SDI represents a highly inefficient way of organizing support for the civilian economy."[48]

Although important, this remark clearly has no practical implications. It might be tempting to underscore the inefficiency with which military programs, under the best of circumstances, manage to favor the development of civilian technologies. At the same time, it must be acknowledged that that has never been their principal objective. Professor Rosenberg's remark nevertheless has the merit of highlighting the weaknesses of a system that has on many counts become anachronistic, since its organization stems from a strong priority—defense—while its results seem to be expected to go in another direction: industrial development.

An Increasingly Criticized Model for Government Intervention

For the past forty years the Department of Defense has played a central role in American scientific and technological development. It gave birth to and nourished a large industrial sector belonging to the sheltered culture. But, as we have seen, it also often stimulated technological development in the exposed culture, at least during the first twenty years after World War II.

Although civilian fallout from military programs has always been haphazard, its existence has given support to the idea that a technology policy aimed at filling the nation's defense needs was globally positive for the economy. The unconvincing studies made during the 1970s on fallout from the space program have not caused this belief to disappear.[49] In the meantime, the geographic concentration of military and space industries has always reminded a certain number of congressmen that the development of these critical defense activities was also essential to the prosperity of their district or state.[50]

The priority given to defense by Ronald Reagan and the enlarged scope of technologies to be developed through government programs in the early 1980s reinforced the image of military technology working to favor the expansion of the rest of the economy. This belief naturally finds adherents among all those working within the sheltered culture. Paradoxically, the coming to power of the free-market advocate Ronald Reagan only served to rejuvenate an old philosophy of government involvement in technological development. In fact this philosophy, whose implicit models are the Manhattan Project and the Apollo program, had never been abandoned. Given its responsibilities for national security, it is up to the government to establish priorities and objectives for technological development and to provide the financial means required. It is up to the scientists and the technologists to do the rest. From Star Wars to the orbiting space station and the hypersonic airplane, the model has found expression, since the early 1980s, in many and varied forms.

In this context, some were understandably misled into believing that DARPA was prepared to meet the Japanese challenge. Was not the Pentagon's advanced-projects division the only place in the federal government where major resources could be mobilized for long-term technological development? What could be more normal, in an administration that at first perceived the Japanese challenge as a problem of technological leadership, than to turn to those who were used to meeting this sort of challenge—the Pentagon's technologists? Moreover, the new DARPA programs could easily be portrayed as fulfilling the United States' ambition to remain the leader in the field of high-speed integrated circuits and intelligent computers, precisely two of the sectors targeted by the Japanese government. It was not particularly noticed that in this race for technological leadership DARPA was primarily concerned with recent progress being made by the Soviet Union, or that DARPA had been organized for that other race. Not noticed either was the growing divergence between the Defense Department's technological objectives and those of the exposed culture. What was taking place was yet another expression of the mirage of technology.

In the light of several realities, however, the time had come to revise the vision of the Pentagon's contributing, through technological development, to the competitive position of American industry.

The first reality was a change in attitude among the industries of the exposed culture toward the Department of Defense. Long indifferent to the extent of the government's spending on military technology, they were now—along with a number of experts,[51] particularly those brought together by the National Academy of Sciences—becoming concerned about the demands made by these programs on an essential resource that is becoming scarce in certain sectors: highly skilled scientists and engineers. Military programs were draining away a growing number of the available specialists in the key sectors of electronics and data processing. In these sectors, the two technical cultures seemed to have become antagonists.[52]

The second reality emerged with the report of the Presidential Commission on Competitiveness, presided over in 1984 by John Young, president of Hewlett-Packard.[53] That commission expressed, for the first time, concern at seeing a disproportionate fraction of the nation's wealth devoted to national security, with respect to another priority that needed even more attention from the government: reinforcing American industrial competitiveness. One commentator noted: "While we're making missiles, the Japanese are making computers." The government's generosity to the sheltered culture was now seen as a threat to the exposed culture. The time for readjustments seemed near at hand, and Ronald Reagan's increasing difficulties with Congress in obtaining approval of the defense budget were the first signs.

One of the reasons for this readjustment was the growing awareness that the Japanese challenge was not only a question of technological leadership and national pride. It was a challenge to the organization of a world that was totally foreign to Pentagon technologies, the world of industrial production and factory organization in the exposed culture. However, few were able to formulate this diagnosis in the early 1980s. American society was engaged in a headlong flight, with which everyone seemed satisfied, toward a tertiary post-industrial economy and the gradual disappearance of its industry.

The emergence of the information society, a consequence of the spread of information technologies, was at the heart of this process.

Questions for the Information Society

8

Historical analyses have shown that technical innovations do not come about uniformly, but in successive waves. Certain periods may see an exceptionally large number of new products and processes; at other times the rhythm of creation seems particularly slow.[1]

Most often, the appearance of one or two fundamental innovations serves as a departure point for a series of new developments and discoveries. After a period of maturation, their combining and cross-fertilization may lead to a mutation of the current "technical system." Science historian Bertrand Gille uses this term to designate the group of technical innovations that contribute to shaping the industrial, economic, and social fabric of a historical period.[2] For example, at the end of the eighteenth century the steam engine and the mechanical loom laid the foundations of the first industrial era. Forty years later came a second wave of innovations, of which the most important were the railroad and cast iron. The third wave, which culminated during the 1890s, laid the basis for a new technical system based on the automobile, steel, electricity, and the telephone. The fourth wave of industrial innovation was the one that occurred between 1940 and 1950. This was the era of the jet airplane, plastic, television, nuclear energy, and mainframe computers.[3] These innovations, born during or just after World War II, are still shaping our industrial landscape and our life styles.

The Fifth Wave

We are now experiencing the beginning of the fifth wave, which involves information technologies. It began during the 1960s with the appearance of the integrated circuit and the microcomputer. It became amplified in the 1980s, a period that saw the progressive spread of these technologies, along with widespread innovation downstream (figure 5). Today we are witnessing a new mutation of the technical system, as these technologies, born of a marriage between data processing and telecommunications, are affecting the lives and the behavior of all economic agents.[4]

The development of the new technologies corresponds to the satisfaction of specific needs: the need for companies to process and transmit increasing masses of information coming from the markets, from customers, and from suppliers spread out over the face of the earth; the needs of researchers facing exponential growth in the production of scientific knowledge; the needs of diverse types of organizations attempting to manage complex systems. It also corresponds to the satisfaction of a potential demand. The need to communicate is growing, along with the appearance of new means of communication, just as a new need for travel arose as soon as modern transportation systems had been created. The railroad, the automobile, and the airplane allowed for the expression of a latent need for travel that was unsuspected in the era of the horse and buggy. The new means of communication allow for the expression of this demand, for they are efficient and inexpensive. The constantly improving performance of electronic components goes hand in hand with a steady decline in the cost of processed or transmitted information.

Two decisive innovations combined to create this new technical system: decentralized computing (particularly microcomputing) and the digitalization of telephone networks, which enabled computers to exchange data and processing results. Telephones and computers are now speaking the same binary language. Numbers, words and pictures can be converted into elementary data "bits," can be transmitted at high speed and in large amounts via light pulses in optical-fiber cables, and can then be organized and sorted using powerful computers. Their simultaneous transmission over

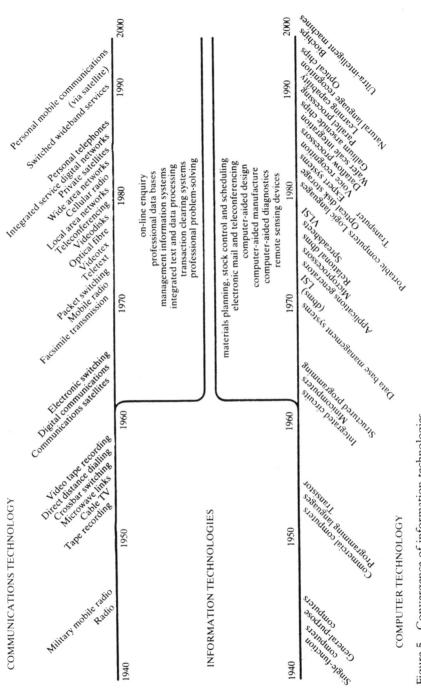

COMMUNICATIONS TECHNOLOGY

Personal mobile communications
(via satellite)
Switched wideband services
Personal telephones
Integrated service digital networks
Private satellites
Wide area networks
Cellular radio
Local area networks
Teleconferencing
Videodisks
Optical fibre
Videotex
Teletext
Packet switching
Mobile radio
Facsimile transmission

Electronic switching
Digital communications
Communications satellites

Video tape recording
Direct distance dialling
Crossbar switching
Microwave links
Cable TV
Tape recording

Military mobile radio
Radio

INFORMATION TECHNOLOGIES

on-line enquiry
professional data bases
management information systems
integrated text and data processing
transaction clearing systems
professional problem-solving

materials planning, stock control and scheduling
electronic mail and teleconferencing
computer-aided design
computer-aided manufacture
computer-aided diagnostics
remote sensing devices

COMPUTER TECHNOLOGY

Ultra-intelligent machines
Biochips
Optical chips
Natural language recognition
Learning capability
Parallel processing
Gallium arsenide chips
Wafer scale integration
Dataflow processors
Voice recognition
Expert systems
VLSI
Optical disk storage
Logic languages
Transputer
Portable computers
Spreadsheets
Relational dbms
Microprocessors
LSI
Applications generators

Data base management systems

Integrated circuits
Minicomputers
Structured programming

Commercial computers
Programming languages
Transistor

Single-function computers
General-purpose computers

Figure 5   Convergence of information technologies.

the same phone line will soon become possible thanks to a new service that is now near the end of testing, the integrated service digital network (ISDN). In fact, as all equipment in service is progressively being digitalized and the number of high-speed links is increased, there is nothing to prevent ISDN from transmitting not only voice and digital data but also pictures, graphs, and facsimiles—in short, everything that can be transformed into elementary data bits. The terminals needed for these services will shortly be available everywhere.[5]

Although we may need to wait a few years before ISDN use becomes widespread, added-value networks (AVN) are already showing what the new forms of communication will look like. They provide for direct communication, usually over high-speed links, between the computers within a company, or within a group concerned with transmitting data and messages or needing access to a service. This type of network is widely used for transferring information and financial transactions between banks. For example, the SWIFT network, created in Europe in 1977 and being continuously modernized, makes it possible to transmit close to a million messages a day between some 2,000 banking establishments and their branches in approximately 50 countries.[6] Payments and transfers of funds between most American banks and foreign correspondents are made via CHIPS (Clearing House Interbank Payment System), a network controlled by a dozen of the largest New York banks, linking 140 American and foreign financial establishments. At the beginning of 1985 CHIPS was handling 200,000 operations daily, representing a total volume of nearly $200 billion.[7] Most of the multinational banks have established private networks between branches, with the center located in the financial market at the hub of their operations. In this way the Bank of Hong Kong and Shanghai is using a private communications network that, via cable and satellite, connects 100 branches in 60 countries, with switching centers in Hong Kong, London, Bahrain, the United States, and Australia.[8] Using international computerized networks, most of the large American banks can offer their industrial customers in various countries a cash-management service that takes account of the dates and places of payments made by customers and suppliers in different currencies. The general use of computers for financial transactions, combined with the broad use of credit cards—and, soon, memory payment cards—is producing a profound transformation of the financial and monetary system. In fact, we are witnessing a transition from an economy based on paper money to one based on electronic money.[9]

In the late 1960s, the number of transactions on the New York Stock Exchange reached 12 million a day. The Exchange needed to be computerized. At that time only the large banks could afford to pay the personnel needed to handle such large numbers of daily transactions. Finally Wall Street's principal operators joined together to create a single firm, the Deposit Trust Company, for centralized stock handling. With that it became possible to transfer stock by purely electronic means, with each broker connected by computer to the central system. The keystone of the system is a network of 200 minicomputers that makes it possible to handle nearly 1,000 transactions per second, including the sale, the purchase, and a record of each transaction. The principal characteristic of this data-processing system, designed by Tandem, is the use of multiple redundancies to guarantee almost absolute security for the user. Faults likely to arise during operation are self-diagnosed and compensated by automatic switching to other parts of the system.[10] In operation since 1975, this system has profoundly changed the way Wall Street works. Operators now have access to information in real time and can react instantly. Wall Street has become a transparent house of glass in which the pulse of the financial markets is registered.[11]

Fast, reliable, and inexpensive control and transmission of large amounts of data has profoundly affected all the financial professions. It also affects many other sectors. Airlines, car rental agencies, insurance companies, hospitals, and chain stores have undergone profound changes thanks to information technologies. Retail business is also affected; the use of bar codes and preprogrammed cash registers allows for simplified automatic management of merchandise flow and inventories.

A White-Collar Revolution

The office is probably the place in the working world that has been most rapidly affected by the spread of information technologies.

Word processors, facsimile machines, video conferencing, and electronic mail have become familiar tools in private firms and public bureaus. The real revolution, however, is the one in microcomputing. Microcomputers are little by little invading the offices of company managers and directors. Easy-to-use software enables accountants, engineers, and financial and marketing managers to make the calculations they need to make, and to continuously update parameters. The widespread use of word-processing software has considerably increased clerical productivity and transformed the relationship between managers and their secretaries. Computers also provide managers with immediate access to the data they need and allow information to be structured in a variety of ways.[12] Connected to one another by local networks, and with the outside via modems and telephone lines, microcomputers have become tools for internal and external company communications. Their number is doubling annually in the United States. In 1987 it reached nearly 30 million units. Services for filling users' needs have seen a parallel explosion. In the area of financial information alone there were 2,400 data banks in the United States in 1986. According to the New York firm Link Resources, the market for electronic information services went from $1.5 billion in 1984 to $3 billion in 1987.[13]

Information Technologies and the Expansion of the Service Economy

The service sector is of growing importance in the American economy. Its contribution to the gross national product rose from 55 percent in 1950 to 71 percent in 1984, while that of the industrial sector declined from 30 percent to 21 percent during the same period. With regard to employment, the change has been more spectacular. The portion of the working population engaged in agriculture has dropped from 40 percent at the beginning of the century to 3 percent in 1985, and the portion working for industry from 25 percent during the 1950s to less than 19 percent today. These declines were counterbalanced by regular growth in the number of tertiary jobs, from 40 percent of the working population in 1950 to 74 percent in 1987.[14]

Nearly three out of four Americans today are working in services. There is a wide variety of services: insurance, banks, travel, education, commerce, health care, telecommunications, hotels, repairs of all types, and services to industrial firms (for example financial and engineering consultants and data-processing service companies).

Information technologies have contributed to this trend in two ways. First, they have done away with barriers to the development of numerous service activities. Manual processing of certain types of information flow had become impossible in the stock market, banks, and insurance companies. The steady improvement in computer performance has now made it possible to handle this nearly limitless flow of data, which is one of the keys to the development of a service economy.[15] Second, the possibility of instantly storing, processing, and transmitting large amounts of information at very low cost has improved the quality of a great number of existing services, such as banking, travel and theater reservations, and consumer information. But more than anything else, the use of information technologies has given birth to a multitude of new services to individuals and to companies. One of the most spectacular examples is the development in France of Minitel, the government-sponsored program providing telephone subscribers with miniature modem-equipped videotext terminals. At the end of 1987 there were more than 6,000 services available to consumers that had been created over the preceding three years. Most of the activities and the jobs created by the development of the new information technologies are in services rather than in industry. The development of hardware—especially computers and telephone switchboards—is increasing the demand for software. In 1960 software represented only 30 percent of the total cost of data-processing systems; today it represents more than 80 percent.[16]

Tasks connected with the handling, processing, and transmission of information involve a growing portion of the employees in the American economy. In 1983 they already accounted for more than one job out of two for the total working population. The portion was 92 percent in the financial services and insurance sector, 62 percent in commerce, and 50 percent in transportation and in the distribution of gas and electricity. Even in industry, tasks associated with information are filling a larger role. They represented

Table 3
Creation of new jobs by sector, 1972–1984.

Activity	Net job creation (in thousands)	Growth rate, 1972–1984 (%)
Health care	2,677	79
Restaurant industry	2,521	70
Wholesale trade	857	53
Banks and financial services	846	54
Food stores	840	30
Social services	667	134
Temporary personnel	613	82
Leisure	445	89
Hotels	418	64
Legal counsel	387	144
Data-processing services	364	325
Real estate	339	46

37 percent of total industrial employment in 1983.[17] The growing dominance of services in the economic and social structure matches the development of what Daniel Bell of Harvard, a visionary of the post-industrial economy, called the "information society."[18]

The Service Sector: An Impetus for Employment?

From 1972 to 1984 the work force of the United States grew by 20.8 million people. This figure represents the net creation of jobs— that is, the difference between new jobs and jobs that had disappeared (of which there were an average of 8 million per year). More than 95 percent of the new jobs were created by the development of services. This contribution has in fact exceeded 100 percent since 1980, since globally the secondary sector lost 1.4 million jobs between 1980 and 1985. Table 3 shows the kinds of activities in which the largest number of jobs were created between 1972 and 1984.[19]

Aside from the government, which created 2.6 million jobs during this period (not shown in the table), health care is in first place, closely followed by restaurants. Most of the new restaurant jobs are in fast-food outlets. The strong growth in this type of job has often been linked—symbolically—to the parallel decrease in industrial jobs. In its race toward services, the United States seems to be progressively replacing industrial jobs with work for the unskilled subproletariat in the fast-food shops. The economist Robert Z. Lawrence asks, in his work on American industrial competitiveness,[20] whether America has become a "nation of hamburger stands."

The Hollowing of Manufacturing Companies

It is true that inadequate industrial efficiency is becoming a major problem for America. Although the labor productivity of American industry is still the highest in the world, this lead is narrowing dangerously. The portion of the GNP produced per salaried employee in the United States in 1985 was $36,900 per year, versus $22,500 in Japan, but the United States has had one of the lowest productivity growth rates over the past 15 years: 0.3 percent per year between 1973 and 1979, and 0.4 percent per year between 1979 and 1984. For the same periods the rates were respectively 3.3 percent and 3.5 percent in Japan.[21] For many observers there is a close correlation between the higher productivity gains of Japanese firms and their success in the American market.[22]

In the early 1980s an additional problem appeared: firms in the cheap-labor countries of the Far East and Southeast Asia began to flood the low end of the American high-technology market with their products. "Clones"—perfect copies—of the IBM PC, made in Taiwan and Singapore, began appearing in the United States in 1982. Their market share grew rapidly. In 1986 the United States imported $1 billion worth of computers from Singapore.[23] American companies adapted to this development either by moving their own production to Southeast Asia or by replacing their least profitable products with imported products.

The relocation of certain factories had actually begun in the 1970s. At that time a large portion of Texas Instruments' integrated circuits were already being produced in Singapore. But the movement accelerated in the 1980s. For example, since 1985 Apple has been producing its Apple II computers in Taiwan. In 1986 AT&T tansferred its consumer-premise-equipment plant from Shreveport to Singapore. In all, American companies imported $70 million from their affiliates abroad during 1986 alone, thus contributing that amount to the American foreign-trade deficit.[24]

For American companies, another way of adapting was to make increasing use of foreign subcontractors and to replace US-manufactured products with imported components or subassemblies from those same cheap-labor countries. This was clearly the option chosen by a large number of firms in the mass-consumer-electronics sector and the computer sector. In many cases these firms went so far as to directly buy finished products for resale under their own brand name. Eastman Kodak buys its video recorders, its video tape, and its midrange photocopying machines in Japan. Apple buys a large portion of the electronic components for its computers in Japan. Even IBM and General Electric, which have always distinguished themselves by the power of their industrial base, are tending to replace large parts of their production with foreign subcontracting. *Business Week* has expressed concern about this: "Outsourcing breaks down manufacturers' traditional vertical structure, in which they make virtually all critical parts, and replaces it with networks of small suppliers. . . . In the long term some experts fear that such fragmented manufacturing operations will merely hasten the hollowing [of industry]."[25]

This evolution is in fact a transposition to high-tech sectors of a phenomenon that is already widespread in traditional industrial sectors. For instance, the American auto companies, in order to maintain their revenues, have been buying an increasing proportion of the vehicles they sell from Japanese manufacturers. According to the Department of Commerce, 17 percent of the cars sold by Detroit in 1988 came off production lines belonging to Toyota, Suzuki, and Mitsubishi. As to the vehicles assembled by American manufacturers, the consultant Arthur Anderson foresees that by 1995, 30 percent of spare parts and subassemblies will be imported, versus 18 percent ten years earlier.[26]

Thus, little by little, American enterprise is calling upon foreign purchases or subcontracting for an increasing portion of its products. In a 1984 book entitled *The Second Industrial Divide*, MIT professors Michael Piore and Charles Sabel forecast a new industrial economic landscape where mass production would be displaced to the Third World, with the industrialized countries retaining only specialized manufacturing.[27] To this new logic of production there is a corresponding new company organization. By progressively giving up large segments of their manufacturing activity, industrial firms are becoming service companies. Though they keep up a certain number of key functions, such as marketing and product design, they are no longer manufacturing anything themselves. They become "hollow corporations," emptied of their manufacturing capacity.[28]

This type of organization has already been adopted by many firms that noticed that, in avoiding manufacturing, they were also doing away with problems tied to managing large numbers of employees. This trend is particularly strong in high-technology industries, where many firms are organized as networks or "solar system" structures, with a large number of external manufacturers—many of them abroad—surrounding a central organization in which the company's vital functions are concentrated. This form of organization is particularly well suited to adjusting to rapid fluctuations in demand, to short product cycles (which require fast shifts to new products), and to rapid technical progress (which accelerates the obsolescence of earlier products)—all of which characteristics are to be found in most markets in the exposed culture.

Sun Microsystems of Mountain View, California, is an example of these "network corporations." This computer manufacturer, one of Silicon Valley's outstanding successes (its sales went from zero to $200 million in less than four years), employs only a few hundred people in manufacturing, out of 3,000 employees. Almost all the components and subassemblies needed to build its computer terminals are purchased from subcontractors, most of them in Southeast Asia and Japan. Sun's only production-related task is to assemble the terminals and check them.[29] With 900 employees in manufacturing out of a total of 2,100, TIE Communications, a Connecticut firm that sells telecommunications equipment, is another example of a hollow company organized as a network.[30]

The Post-Industrial Economy in Question

Many observers of the American industrial scene view the industrial firms' race to services as part of an inevitable evolution. Why, after all, manufacture something that can be produced elsewhere more efficiently and at lower cost? At the close of the twentieth century, high technology and services are obvious comparative advantages that America has in the international division of labor. But Berkeley professors Stephen Cohen and John Zysman, in their book *The Myth of the Post-Industrial Economy*, reply that this is an illusion.[31] The expansion of a service-based economy is impossible unless a solid industrial base also exists, they say. By letting its industries move abroad, the United States has put itself in danger of compromising its future. Their conclusion rests on several arguments.

The first involves an accounting impossibility: an excess in the trade balance for services will never be able to compensate for a deficit in manufactured products. Therefore, the United States must continue producing manufactured goods in order to ensure its prosperity.

The second argument of Cohen and Zysman stems from a common-sense observation. The transition that occurred during the first half of this century from a predominantly agricultural economy to an industrial one was not accompanied by any abandonment of agricultural production. Indeed, it meant the stabilization or even the growth of added value in the agricultural sector. But that was obtained at the price of a considerable increase in productivity, which freed for industry the labor that had once been busy in the fields. Similarly, today it would be a mistake to believe that the transition to a service-based economy can or should be accomplished at the expense of industrial production. The postindustrial society means not a society without industry but a society that is more efficient in producing industrial goods and services.

The third argument underlines the fact that American firms are in danger of losing their qualification to produce certain high-added-value services, such as engineering, if they lose touch with industrial production. Cohen and Zysman remark that the large American engineering firms are increasingly being outdistanced by their Asian competitors in the design and implementation of production units in the steel and petrochemical sectors—markets in which the US position is eroding.[32] This argument is linked to a question asked by MIT professor Lester Thurow: "If American firms lose their industrial base, how will their engineers and scientists keep abreast to sell their services abroad?"[33] In order to take advantage of the learning process, a large portion of technological innovations and breakthroughs must be generated at the core of the industrial process, not in laboratories. A firm that loses contact with manufacturing is in danger of rapidly losing its capacity to innovate. No longer concerned with the need to lower production costs, the firm that subcontracts production is in serious danger of losing its capacity to identify the innovations needed for the next generation of products.

Thus, it is not certain that any American firm, even if it wanted to do so, could today manufacture compact-disc players at a competitive price. Little by little, by abandoning production in almost the entire hi-fi sector, American companies have entirely lost the capacity to innovate in this field.

Akio Morita, Sony's co-founder and an attentive observer of the American scene, summarizes this evolution in the following terms: "American companies have either shifted output to low-wage countries or come to buy parts and assembled products from countries like Japan that can make quality products at low prices: the result is a hollowing of American industry. The US is abandoning its status as an industrial power."[34]

Mirage of Technology or Mirage of the Information Society?

Wall Street plays a central role in the developmental dynamics of American high technology. It is the perspective, however distant, of introducing their company on the stock market that motivates the majority of new entrepreneurs, as well as the venture capitalists financing the development of their firms. This hope is often shared by the employees of these new companies. They accept having to work hard and sometimes having to tighten their belts to help the new company take off, and they receive in exchange a few shares of stock. This is a universal rule in Silicon Valley. This stock will—perhaps—turn into gold if the project succeeds. Always cited is the fact that a hundred or so of Apple's employees became millionaires the day the company's stock was quoted on the stock market. Even certain established firms are relatively generous with their employees in this regard. For example, Genetech allowed each of its employees to acquire 100 shares for a modest price the day the company received its first marketing license from the Food and Drug Administration.[35] In Silicon Valley stock is sometimes used like cash. The story is told of broke entrepreneurs who continue to pay their rent with stock of the company in which they had invested all their resources.[36] Owners generally accept it. But nothing could be less certain than the future value of certain shares, particularly if they belong to "high-tech" companies.

Yet everyone is caught up in this gambling atmosphere that characterizes the world of high technology. "The greatest profit goes with the greatest risk," says Morton Davis, president of D. H. Blair & Company, a New York investment consulting firm. Davis has a liking for what stock traders call "high-flying stock," meaning shares in companies with record growth. It is a dangerous exercise, but the rewards can be great. It is not surprising that the promised land for Morton Davis is high technology. In early 1987 he was advising his clients to invest in the AT&E Corporation, a firm that had developed technology making it possible to miniaturize paging receivers in the form of a wristwatch that would signal a call and display the caller's number. Between January and April 1987, AT&E stock rose from $8 to $32. At that moment, the company's value was estimated at $290 million, even though sales had never exceeded $1 million. The Blair Company did not hesitate to forecast AT&E sales of more than $1 billion by 1992. But as of May 1 the horizon began to darken. Problems in perfecting the product frightened investors, and the stock fell to $24.[37]

Biotechnology is a high-risk sector if ever there was one. Its unique characteristic is to have given birth in the late 1970s to several hundred new firms, without any of their sales potential having yet materialized.[38] There are good explanations for this, however. Progress in genetic-engineering research has been slower than predicted. In addition, the concern raised by the introduction of genetically modified organisms into the ecosystem has incited regulatory authorities to be cautious. Long and complex administrative procedures have delayed many experiments. The result has been strong natural selection among the firms in line to collect the benefits of the biorevolution. But the most persevering investors should be rewarded eventually. Although they have been substantially deferred, the promises of this biorevolution have remained intact. In 1995, sales of products issuing from biotechnology could reach $10 billion.[39] The problem is to identify, in the throng of new companies in this sector, which ones will be able to hold out to the end, to gather the expected manna. This explains the two paradoxes in Wall Street's behavior with respect to biotechnology.

The first paradox stems from the divergence between the stocks' value on the market and the companies' sales. Although these stocks have been quoted on the stock market for years, their value has relatively little relationship with the companies' current ability to produce and sell a product. For example, Cetus—one of the oldest of the new companies in this sector, created in the 1970s—by 1987 had yet to market a single product resulting from genetic engineering. The first results of its research, Interlukine-2 and Beta Interferon, were expected to be marketed in 1988. The company's 1987 profits—$1.2 million—came essentially from financial operations.[40] In fact Cetus has a war chest: about $90 million in cash and a $200 million credit line, provided by investors convinced that the company represents a reasonable gamble for the future. However, this has not protected Cetus from the second paradox of the biotechnology stock market: extreme price volatility. In their search for reference points to enable them to estimate the "true value" of companies whose results are still only potential, investors are sensitive to the slightest bit of information that can help them pick the future champions. Thus, the announcement in late December 1985 by Dr. Rosenberg of the National Cancer Institute that promising results had been obtained using Interlukine-2 to treat certain types of cancer boosted Cetus' stock. After having fallen below $9 in mid-1985, its price exceeded $28 in January 1986. But the information urging investors to gamble on falling stocks is just as abundant as that which could incite them to gamble on others going in the opposite direction. Viratek is a firm that suffered from such a situation. It developed Ribavirin, a drug that in 1986 showed promise for treating certain forms of AIDS. As soon as it was known, this news produced a dizzying escalation of the company's stock, which in a few weeks rose from $10 to $98. At that moment the company's worth was estimated at $788 million. Unhappily, clinical experiments did not confirm the hopes founded on Ribavirin. In September 1987 the Food and Drug Administration refused to approve the drug for marketing. By October, Viratek's stock had plummeted to $16.

As the *New York Times* commented, "Investors are discovering that certain high performance stocks only have a story—and nothing else—underlying their performance."[41] Their advocates describe them as the companies of tomorrow, the new Apple, Xerox, or Syntex; their critics call them "puff stocks" to underscore the fact that they are based more on illusions than on reality. It is true that there is always some bluffing in predictions of the performance of products that depend on technical developments that have yet to be mastered. A scientist or an engineer promoting his idea is generally full of certainties. Insofar as his promoter has decided to assume the financial risk, he should make an effort to share them. In "high tech," results often count less than potential. It is the development of this potential, along with an ability to market it, that will determine the subsequent value of the stock of a high-technology firm.

Naturally, the problem for investors is to judge how realistic the objectives are, to distinguish between real and imaginary potential. For uncertainty is the main feature of this game. This is why the slightest rumor relating to a company's results or progress can produce truly disproportionate reactions. Copytele, Inc., experienced this in 1987. This company had based its development on the creation of a high-performance liquid-crystal display. There are many applications for such a device; including TV receivers and computers. The market is gigantic. If Copytele is successful, some of its active supporters estimate its future value at a billion dollars. But this is a high-risk sector. Many other companies are also striving to develop similar products to conquer this market. In April 1987, Copytele shares, which were being quoted at $10, suddenly flared to $25. The company's value followed the same rising movement, reaching $185 million although it had yet to sell anything. The rumor of an imminent agreement with a large Japanese TV manufacturer was the reason. But it was only a rumor, and a few weeks later Copytele stock plunged to $13.[42]

Paradoxically, the volatility of stock prices seems to increase as investors' information increases—a phenomenon not limited to high technology. The multiplication of personal computers has produced a proliferation of "decision-making" software that enables the man in the street to be just as well "connected" as the best Wall Street analysts. All he needs is a PC, a modem, and a phone line. There are two major categories of investment decision-making software: "fundamental analysis programs" (for identifying undervalued companies with good long-term prospects) and "technical programs" (which provide a number of complex indicators reflecting market changes, to help the investor reap profits from rapid price fluctuations). Subscribing to the simplest of these services may cost $275 a year; the most expensive ones are $1,500 a year. In 1987 Criterion Software began marketing a new and very comprehensive technical program for under $700. On October 19, 1987, happy Criterion customers suddenly saw a warning appear on their screens. On a scale of 1 to 100, with 100 meaning "Sell everything at once," the program was displaying the figure 99.[43] Wall Street had just taken the biggest plunge in its history.

On that day John D. Phelan, Jr., president of the New York Stock Exchange, reached his office at 7:00 A.M. The preceding Friday had been disastrous. The Dow-Jones average had lost 108 points. Today he was expecting the worst. Even before the opening bell, his computer terminal was showing a tendency indicator based on a survey of the first sell orders. IBM shares, which had fallen to $134 Friday afternoon, were being quoted at no more than $125 at opening this morning. The first hour was even worse than predicted. By 10:30 the Dow-Jones had already lost 104 points. Panic sales were also running through the Chicago Mercantile Exchange, the centerpiece of the futures market. Standard & Poor's 500 Index, the principal quotation for futures and options, was following the same descending spiral. But in New York a special concern was on the minds of all those responsible for the operation of the swarming hive that is Wall Street: would the computer system hold up? Since the start of business, John Phelan's associates had been in touch with the major stockbrokers. They were deluged with customer orders, but were managing to cope with the situation. The New York Stock Exchange's computer system had been designed to handle a maximum of 450 million transactions a day. By 1:00 P.M., 330 million shares had already changed hands. The system was becoming clogged. The tapes recording the transactions being processed by the computers were running 90 minutes behind.[44]

But the worst was yet to come. As the market continued to plunge, operators of pension funds and insurance companies, attempting to compensate for their losses on the stock market, began to sell stock index futures on a massive scale. This involves a technique known as portfolio insurance, which is most often run with the aid of computer programs that automatically initiate trade in multimillion-dollar lots, with no human intervention.[45] These sales accelerated the fall in Standard & Poor's 500, which then began to drop faster than the Dow-Jones. These cheaper futures, in turn, induced massive sales by the many operators using what is known as the index arbitrage technique, another computerized trading method. Here the goal is to take advantage of the difference— minimal but constant—between the prices of stocks and the prices of futures. When futures fall below stocks, stocks are sold. When stocks fall below futures, the opposite is done. Since the differences are generally very slim, huge lots must be traded before a profit is shown. Hence, early in the afternoon of October 19, program traders using index arbitrage jumped in to take advantage of the opportunities created by portfolio insurers' sales. By selling huge amounts of stock, they accelerated the downward spiral. At 4:00 P.M. Wall Street closed with the Dow down 508 points from the preceding Friday—22.6 percent of its value. There had been more than 600 million transactions. But John Phelan had good reason to be satisfied on at least one aspect of this disaster: although the overloaded automatic quotation system had come to a standstill several times[46] (the longest halt lasted 13 minutes in the early afternoon), his computer system had stood up, and the Exchange had been able to handle the situation.[47]

Lessons from the Crash

Many analyses have been made—and are still being made—of the causes of the 1987 crash. Whose fault was it?

Was it the fault of computerized transactions, which projected the picture of a society dominated by computer decisions and no longer cognizant of where it is going? Was it the information society itself, which had gradually lost its landmarks to such a degree that it allowed what has been portrayed as a "financial bubble" to expand out of all proportion?

In a few short years, data-processing tools have changed the face of the world's financial markets. They have made it possible to create new products (notably the financial futures on the Chicago Mercantile Exchange) that represent another step in abstraction and speculation, imperceptibly bringing the rules of the financial markets closer to those of a gigantic casino. They have also led to computerized decision-making techniques that substitute machine-made decisions for those made by investors.

Not only do computers mechanize transactions and transmit and process a lot of numbers; they also make possible the rapid manipulation of large amounts of data. In index arbitrage, for example, the computer alone determines a strategy and implements it, minute by minute, buying and selling millions of stock shares.

Of course we are still far from HAL, the computer master of the spaceship in *2001: A Space Odyssey*. But the autonomy of these machines is a cause of concern. In the *New York Times* of October 19, 1986—exactly one year before the crash—David Sanger noted that many financial experts—to say nothing of the frightened small investors, were accusing computers of having created a machine-managed market. At that time, computers were already suspected of having caused the 86-point drop in the Dow-Jones the preceding September 11. Many specialists, the *Times* noted, felt that computer-aided sales techniques were responsible for creating gigantic and chaotic market fluctuations.[48]

Carried along by ever-wilder speculation exacerbated by instant financial information being updated by worldwide computer networks, the world's large financial centers—following the example of Wall Street—have little by little become disconnected from the real economy. Already in 1986 numerous observers were speaking about the dangerously increasing divergence between euphoric stock markets, moving along a sort of ascending spiral, and the reality of the corporate world. In the above-mentioned *Times* article, David Sanger added that, by making it possible to instantly analyze a large number of figures and to take advantage of short-term market variations, new software and machines had drawn institutional investors into trading operations that had little to do with the companies' fundamental value or their economic prospects.

The profits accumulated on the magnetic tapes of the "computer traders" had less and less to do with the success or failure of the firms themselves. Indeed, the existence of these companies was less and less important. In a post-industrial society, appearance probably counted more than reality. Moreover, was there anything real about the fortunes being amassed by the computers?

The two chief consequences of the 1987 crash were that it revealed the mirage of the information society and that is brought about a salutary return to reality.

Japan, Europe, and the Two Cultures

9

Figure 6 presents a summary of the major findings discussed in the preceding chapters. It illustrates a dramatic decline in the relative competitiveness of American high-tech industries—a decline that benefits Japan in the information and communications sectors and Europe in aerospace and armaments. For each of these industries, the figure shows the historical change in the relative trade balance between each country and the rest of the world.[1]

These striking results strongly suggest a need to better understand the reasons for the strength of Japan and Europe. Where did Japan get such an irresistible comparative advantage in the exposed culture? What can explain Europe's international breakthroughs in sheltered industries?

In order to try to answer these questions, we need to understand the logic of technological innovation in Europe and Japan.

For this purpose I shall apply the model I used to analyze the organization of high-technology industries in the United States. As one might suspect, the same dual pattern of technology can be observed in other countries. In most industrialized countries, technological development results from a complex blend of exposed and sheltered cultures. But although in the United States each of the cultures influences a large segment of industry, in most other countries one culture or the other is dominant. This may be the case for various reasons: a scientific and technical heritage, a tradition of government intervention, a particular organization of industry, or limited financial resources. It may also sometimes result

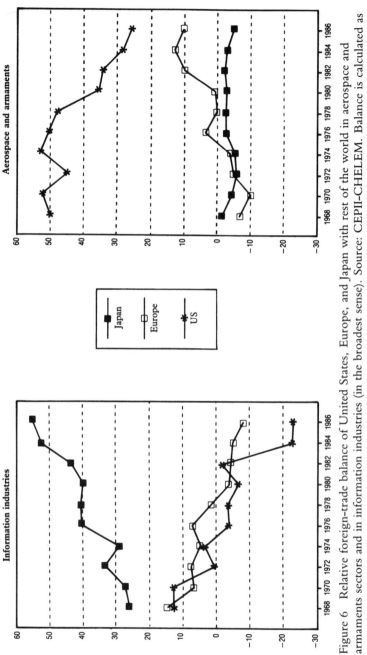

Figure 6 Relative foreign-trade balance of United States, Europe, and Japan with rest of the world in aerospace and armaments sectors and in information industries (in the broadest sense). Source: CEPII-CHELEM. Balance is calculated as $100 \times (X - M)/W$, where X represents exports, M imports, and W world trade by type of product.

from deliberate strategic governmental choice. Whatever the reasons, however, the prevalence of one culture strongly affects the comparative advantage of a country's companies in international competition. The examples of Japan and Europe will show us how this comes about.

Japan: Archetype of the Exposed Culture

The spectacular growth in Japan's share of world trade in some high-technology products is due in part to historical circumstances, but also in part to a deliberate choice by the Japanese government. Japan is one of the few countries whose leaders have been able to articulate a long-term strategy for industrial development, including an important international dimension, and to implement it via close cooperation between industry and government. The example of semiconductors will highlight the nature of this relationship and the true reasons for Japan's superiority in this field. It will also provide the background information we need to characterize the Japanese technical culture.

The Japanese strategy for penetrating the world's semiconductor markets only repeats the well-rehearsed scenario that enabled Japan to emerge during the 1960s as an industrial power with worldwide ambitions in sectors such as steel, shipbuilding, and automobiles. In the years after World War II, the Ministry of International Trade and Industry (MITI) decided to concentrate governmental and industrial resources in a limited number of specific sectors—sectors in which rapid growth in world demand could be anticipated, and in which there was strong demand in Japan as well. Conquest of the domestic market was mandatory as a first step in the MITI strategy. It was facilitated by trade barriers to prevent foreign competition. Unable to enter the Japanese market, foreign firms generally had no choice but to accept licensing agreements proposed by Japanese firms for exploiting their patents in Japan.[2] Such access to foreign know-how was crucial for Japanese firms, whose first objective was to duplicate the best of what was being produced abroad. The next step, and a precondition for the conquest of international markets, was to lower costs and use the economies of scale associated with mass production. Consolidating and broadening the know-how thus accumulated would enable Japanese firms to diversify and extend their influence wherever they had been able to gain a foothold. This is why MITI encourages R&D in industry and contributes to its financing.[3]

Gradually the Japanese government moved from targeting traditional industries toward targeting high technology. After automobiles and consumer electronics, the semiconductor sector was selected as a target. The first government decisions in this field go back to 1971. MITI engineered the creation of three pairs of companies—NEC and Toshiba, Oki and Mitsubishi, Fujitsu and Hitachi—to encourage cooperation in semiconductor research between rival firms. The objective was to create several industrial associations, each developing generic technologies suited to a particular market niche. The firms cooperated in research and then became competitors in manufacturing the products.[4] However, MITI coordinated the competition, in order to encourage the companies to specialize in complementary sectors. This aspect was essential, for the Japanese market for integrated circuits was still in its infancy in 1975. This preliminary setting for the semiconductor industry had been prepared to a great extent by the Japanese firms themselves during the preceding period. Beginning in the late 1960s, they made massive license acquisitions from American semiconductor firms. This enabled them to produce their first 4K DRAM memories in 1974 and to undertake the reconquest of their internal market for integrated circuits, which until then had been supplied by imports from the United States.

The Japanese had two decisive trump cards to play in doing this. The first was their ability to master manufacturing technologies. This enabled them to descend the "learning curves"—which, as we have seen, are important in the integrated-circuits industry—more efficiently than their American competitors. The second trump card was a capacity to upgrade the technology toward larger-scale integration and greater complexity, thanks to a vigorous R&D effort.[5] This upgrading enabled Japanese firms to produce successive generations of integrated circuits at minimal cost, to the point of complete domination in some product lines. Another advantage Japanese semiconductor firms have over their American counterparts comes from their strong vertical integration. In contrast with

most American manufacturers in this area, semiconductor production represents only a modest fraction of turnover for the Japanese
firms involved: 20 percent for Hitachi and Toshiba, 30 percent for
NEC.[6] The result is decreased financial fragility, an important asset
when there is a need to face large demand fluctuations (a frequent
phenomenon in this highly cyclic industry). Moreover, the company can afford to fight to preserve a market share when over
production and price collapse occur, as they did in 1985. Such
vertical integration also helps Japanese companies find the huge
resources needed for investments to develop new generations of
products.[7] This advantage turned out to be decisive in the harsh
competition that started at the beginning of the 1980s between
American and Japanese semiconductor manufacturers. Japanese
firms were able to increase their productive investments by 72
percent in 1983 and by 104 percent in 1984, becoming worldwide
leaders.[8]

The Japanese government's role in this was essential, but was
also very limited. MITI provides the strategic framework for the
companies' action by coordinating their objectives during the critical development phase; it encourages them to share the R&D effort
and participate in its financing, but it does not interfere in their
technological or industrial choices. The structure of the industry is
clearly the result of intense competition between the companies for
market shares. But at the same time, they have the benefit of solid
protection against foreign competition. Japan imports only the
products it does not know how to make itself.[9]

The above description characterizes most of the electronics and
data-processing sectors. Although there are some variations, MITI's
intervention philosophy is always the same. First of all, MITI shows
the road to take and establishes long-term global industrial targets—
the first "plan," for promoting the mechanical industry, was
launched in 1956, the one for consumer electronics in 1957. Next,
it stimulates growth and protects young industries: MITI has
become, in a sense, like a gardener, raising industries in green-
houses. Domestic competition and solid protection from outside
are the two basic rules of Japanese industrial organization. In his
latest book on Japan, Christian Sautter notes that "in the overheated
greenhouse a pitiless struggle is under way to conquer a market of
100 million inhabitants." The third of MITI's tasks is precisely to

oversee this competition, in order to lessen it should it become too dangerous.[10] Contrary to the beliefs of many in Europe and the United States, MITI's active role in fostering the development of the Japanese electronics industry has little in common with past government involvements aimed at the development of large technological projects in Great Britain, France, and the United States.

It should not, however, be concluded that the sheltered technical culture does not exist in Japan. For instance, the high-speed "bullet train" illustrates a certain tradition of large government programs calling for research aimed at technical performance. Japan's telecommunications sector, which was the source of a flood of government purchases until it was privatized in 1985, also displays many of the characteristics of the sheltered culture. In a comparison with the American system, however, it shows important differences.

The first of these differences is that in Japan there is no equivalent to Western Electric, the Bell System's exclusive manufacturer before 1983. On the contrary, NTT, the national telecommunications company, operates in close relationship with a large number of suppliers.[11] It provides them with financial support for research, but it calls for competitive bidding for the supply of equipment. This structure encourages innovation, particularly in telephone sets and other consumer-premise equipment. It favors the creation of low-cost, bottom-of-the-line products, which are also the easiest to export. This is one of the reasons for the Japanese success in the American market for consumer-premise equipment.

The second difference comes from the small relative weight of government procurement—or similar markets—in the sales made by the largest telecommunications manufacturers in Japan.[12] Central switching equipment represents only 12 percent of sales by NEC, which is NTT's chief supplier. Telecommunications products as a whole represent only 28 percent of this highly diversified giant company's $12 billion revenue. Two-thirds of NEC's activity comes from data processing, electronic components, and consumer electronics. As a leading industrial group in the most sheltered telecommunications equipment market, NEC nevertheless has the dominant part of its activity within the exposed culture. Furthermore, when we examine the activities of the ten largest Japanese industrial groups in this sector, we see that the essential part of their revenue—and often all of it—comes from consumer electronics, semiconductors, and data processing. Firms in the exposed culture largely dominate the communications industries in Japan.[13]

Another approach to understanding the reasons for Japan's specialization and superiority in so many sectors of the exposed culture is to look at the reasons for the weakness of the sheltered culture in Japan.

One of these is unquestionably the weak scientific tradition in that country. The sheltered culture can best develop when it is grounded in a solid tradition of fundamental research. One would search in vain for a Japanese equivalent of the Bell Labs.[14] In 1965, Japan's national expenditure for research was 4 percent of the American expenditure. It was only during the 1970s that Japan caught up in this domain. Even then the effort was essentially directed toward applied research and product development, not basic science. Even today the Land of the Rising Sun does not shine by the number of its Nobel Prize winners, and the American government never misses an opportunity to point out Japan's minor contribution to general worldwide progress in scientific knowledge. Today Japan's national R&D expenditure is 65 percent of that of the United States.[15] Until recently, technological development in Japanese firms was almost entirely based on the acquisition of patents from abroad, particularly from the United States. Today the Japanese technological balance—the difference between the sales and purchases of patents abroad—has reached equilibrium, but it showed a deficit of more than $800 million in 1973.[16] Foreign acquisition of know-how is an approach that is particularly well suited to companies in the exposed culture. Market needs command the selection of the technologies they have to develop and hence the patents they need to acquire. Assuming that this purchase is possible, there is no need to make great inventions to manufacture and sell products with high technical content. However, it is essential to know how to adapt the technology to manufacture these products efficiently—an art in which the Japanese have become masters. The Americans invented the transistor, but calculators stamped "made in Japan" have conquered the planet.

A second hypothesis for explaining the limited extent of the sheltered culture in Japan is tied to historical heritage. Although prior to World War II Japan had built the world's most powerful navy and air force, in 1945 it found itself prohibited from maintaining an industrial sector tied to defense activities. The Mitsubishi and Kawasaki factories that had built Zero-Zeka fighters and Lily bombers during the war had to be converted to civilian activities.[17] Today Japan still devotes only 2 percent of its public finances to military R&D, versus more than 70 percent in the United States. This explains the relative weakness, until recently, of the Japanese aerospace industry, as well as of all other industrial sectors related to defense activities. In the sheltered culture, know-how disappears rapidly if it is not applied to concrete projects. This, along with the weakness of the domestic Japanese market for civilian aircraft, probably explains the unfruitful endeavors of Japanese firms in this sheltered sector. Nathan Rosenberg of Stanford recalls his discussions with Mitsubishi executives about the reasons for the weakness of the Japanese aircraft industry. They admitted that Japan lacked experience in pulling together the numerous technologies needed to make a modern jetliner, the characteristics of which must, moreover, be discussed with potential customers, each of whom wants a custom or semi-custom design to fill specific needs.[18] This probably temporary handicap for Japan's sheltered culture is seen by some as a very modest disadvantage in comparison with the clear relative advantage the Japanese were able to develop in spite of the restrictions to which they were subjected. As one Japanese industrialist has remarked: "Our impossibility of creating a weapons industry has turned out to be an important advantage. We have had to use imagination, and concentrate on the new growth industries such as automobiles and electronics."[19]

The third reason is both strategic and cultural. Japan has often been portrayed as a country with two assets favoring the creation of a great national project: strong nationalism and a quest for consensus at every level of its political and social organization. In the 1930s these forces were used in the service of an expansionist policy and military conquest. Many in the West see similar Japanese behavior today, transposed to the economic arena.[20] The conquest of world markets and the mobilization of the entire nation's energy to attain this objective are seen as Japan's indirect revenge on the victors of 1945. In this light, Japan's relative specialization in exposed-culture industries would be the result neither of chance nor of historical heritage—the country proved before the war that it was capable of being a major player in aeronautics and armaments—but of necessity. Today Japan has chosen to specialize in the industrial sectors in which it can hope to conquer world markets, given the restrictions imposed at the end of the war.[21] Furthermore, it has selected the industries of the future: by the year 2000, the world market for information and communications technologies should be above $1 trillion per year.

For large Japanese firms that are highly diversified, this aggressive export strategy has been favored by the strong protection they enjoy at home: most Japanese domestic markets are efficiently protected from foreign competition. Such protection often results from technical as well as cultural factors, including the effect of government policies combined with the nationalistic attitude of Japanese consumers. The result has been the emergence of a dual economy in Japan. One one hand, we can find export-oriented industries fostered by MITI, organized around highly efficient companies that are specialized in the technologies of the exposed culture and that have fully adopted the behavior and rules of this technical culture. On the other, we find industries specialized in filling the needs of domestic markets totally insulated from foreign competition.[22] These markets (in small part for high-tech equipment but predominantly for low-tech products, such as appliances and food) are often dominated by monopolies or oligopolies that do not seem to be models of efficiency, to say the least. This closure of Japanese markets to foreign goods has become a major source of conflict between the Japanese and American governments. Indeed, while becoming the top exporting nation in the world, Japan has thus far avoided reciprocity in its exchanges—a strategy that has made many observers identify the country as an unfair player in international competition.[23]

Through a mixture of free competition and state intervention, in Japan the exposed technical culture has found a promised land. Although this culture thrives here under very special conditions, it has nevertheless kept its most essential feature: it characterizes companies involved in the production of standardized products with a high added value, distributed to wide range of customers in highly competitive markets.

The Japanese example illustrates an important aspect of the distinction that I have made between the two cultures. MITI's role in fostering Japan's exposed culture industries confirms that what sets the two cultures apart is not the presence or absence of government intervention, but rather the structure of the demand and the industrial organization adapted to it. It also shows that there are forms of government intervention that are compatible with the rules of the exposed culture, and that contribute to stimulating instead of smothering the competitive capacity of firms. American manufacturers of semiconductor, and some in other industries have correctly perceived that the Japanese threat does not come only from Japanese firms' access to the most advanced scientific and technical knowledge; it is also due to their remarkable adaptation to the rules of industrial efficiency that exist in the exposed culture.

A very different situation is to be found in Europe, where the juxtaposition of different political, scientific, and industrial traditions has led to a different balance between the two cultures.

The Difficult First Steps toward a Technological Europe

We can say that the history of technological Europe began in 1957. That year saw not only the birth of the Common Market but also that of Euratom, an organization created to promote European development of civilian nuclear energy.

Since 1946 Great Britain and France had given high priority to nuclear power. The two countries had established powerful public research instruments—the Commissariat à l'Energie Atomique (CEA) and the Atomic Energy Authority—which drew the best talent into this new field. Archetypes of the sheltered culture, the nuclear research centers were among the temples of European physics. Saclay in France and Harwell in Great Britain were in a way the "Bell Labs of Europe," by virtue of their atmosphere, the quality of their researchers, and the magnitude of their resources.[24] There was intense rivalry, however, between these centers. At stake, in addition to important fundamental discoveries, was the mastery of civilian atomic technologies, seen as one of the keys to future economic development. Nuclear energy was expected to replace coal for electrical generation in a Europe that was woefully short of hydrocarbon resources. Great Britain produced its first nuclear kilowatt-hours in 1956, at its Calder Hall reactor. The following year, the CEA symbolically connected its second prototype reactor, at Marcoule, to the French electrical network. Other European countries, with smaller resources, also took an interest in nuclear energy. In 1958 Italy ordered its first reactor from General Electric. Germany and Belgium soon followed suit.

In this context, the Euratom treaty raised great expectations. Everyone agreed that developing nuclear technologies would be costly, and that in addition to building power plants it was necessary to master complex technologies to produce nuclear fuels and to handle the highly radioactive wastes. Euratom was intended to make it possible to pool European resources for these purposes. This enthusiastic vision was, unfortunately, not realized. A "technological war" poisoned relations among the major European countries in the realm of nuclear cooperation.

France and Great Britain developed two very different varieties of gas-graphite reactors—using carbonic gas as the coolant and graphite as the moderator—whereas the United States developed ordinary water-cooled reactors and Canada used heavy water. The CEA, which felt, not unreasonably, that France was ahead of the other Euratom countries technically, hoped to impose the French technology on its neighbors. This French imperialism irritated the other participants—particularly Germany, which was inclined toward the American-style light-water reactor. Very soon Euratom meetings became battlegrounds. The French representative would almost systematically obstruct proceedings once it became clear that the other member countries would not be using the French technology. The common research programs that enabled the organization to save face were only a facade. The Euratom member states watched carefully to see that research carried out in common would not touch on any sensitive technology being developed secretly under national programs.[25]

The American technology soon came to dominate a divided Europe. France—the next-to-last country to rally to it (Great Britain followed after a few years)—capitulated in 1969 after intense conflict between proponents and critics of the gas-graphite reactor. EDF, the French public utility, would henceforth buy light-water reactors built in France under Westinghouse license by Framatome, a subsidiary of the Creusot-Loire group. In Germany, light-water reactors would be built under American license by KWU, a Siemens affiliate. Each of these large firms had a monopoly for the domestic market in its country. In this sector, which is particularly representative of the European sheltered culture, the rule was "each to his own." Each government reserved its own markets for its own national champion, so that it could sharpen its sword for the real battle: the international market. It became virtually mandatory to sell abroad to recover at least some of the heavy investments made to develop the nuclear reactors and their fuel cycle. This hope of exporting was one of the reasons the French eventually adopted the American technology. The European countries that had no nuclear industry—Switzerland, Italy, Spain, and Belgium—became the object of fierce battles among the large European and American reactor manufacturers.

The petroleum shock of 1974 raised the stakes in these industrial contests. In most Western countries, accelerated nuclear programs appeared to be a means of responding to OPEC, the new cartel of petroleum-producing countries. For countries with small hydrocarbon resources, the development of nuclear energy seemed to be one of the most effective ways of reducing petroleum imports. Forecasts of worldwide nuclear power soared to more than a thousand 1,000–megawatt reactors by the turn of the century.[26] Intense competition began among the giants of the European nuclear industry. No fewer than eight European companies were involved, including Framatome and CGE in France, Siemens-KWU and MBB in Germany, Brown Boveri in Switzerland, and ASEA in Sweden.

This competition would prove to have been in vain a few years later. The expected manna did not appear. Orders for reactors slowed, and stopped entirely in Germany and the United States, toward the end of the 1970s. The causes were grossly overestimated electrical demand, a reversal of the petroleum-price situation, and local opposition to nuclear power by environmentalist groups with awesome political clout.[27] The developing countries soon encountered similar phenomena and reactor exports dried up.

Framatome was able to remain active while most of its competitors were without orders. At the end of 1986 the proportion of nucleargenerated electricity was 70 percent in France, 29 percent in Germany, 25 percent in Japan, and only 17 percent in the United States.[28] The French manufacturer seemed to be one of the rare survivors in a fading industry. Only by integrating their nuclear branch within diversified industrial groups were certain firms able to survive as nuclear suppliers. But the future is not bright for the French manufacturer. France today possesses an excess of installed nuclear capacity, and the prospects for export are extremely dim.

In another sector—aircraft—the convergence of European interests at the end of the 1950s was also clear. France and Great Britain were the European countries with the most powerful and innovative aircraft industries. However, they were spread out among several manufacturers, whereas in the United States there were two giants (Lockheed and Douglas). There was an obvious interest in establishing Franco-British synergy as a means of reaching "critical mass," an indispensable condition for entering world markets. However, the French and the British firms were strong competitors. Both Britains's de Havilland, with its Comet, and France's Sud Aviation, with the Caravelle, had their eyes on foreign markets. The idea of cooperating on a new generation of aircraft came up against the rivalries of the moment and the quarrels among the two companies engineers as to which group had the best technical solutions.

Another area for agreement seemed possible, however. Why not build the plane of the future—a supersonic transport—together?

The Concorde program, born in 1960 in the form of industrial cooperation between Nord Aviation and British Aircraft, was the result of a political objective shared by the French and British governments. The basic idea was relatively simple: to transpose to the civilian sector the experience accumulated in both countries in the field of supersonic military aircraft. Although this initial hypothesis was quickly to prove false (the know-how needed to build a fighter able to fly at Mach 2 for short periods differs from that needed to build a commercial aircraft to cruise at that speed), the program was continued, with successive reevaluations of its cost. Started on the basis of totally wrong cost projections and future market estimates, Concorde was threatened several times—in particular by the British, who considered withdrawing. But it survived these hurdles and soon became the symbol of a Europe attempting to catch up technologically with the United States.[29]

At the time, the United States was refusing to supply France with the computers needed for its deterrent nuclear force, and Gaullist France's large technological programs looked like national fist-shaking. Even though the abandonment of the American SST project in 1971 cast doubts on the validity of the Concorde project, the self-confident engineers directing the Franco-British program saw it as proof of the Concorde's superiority. The Franco-British project therefore continued. Within the European sheltered culture, programs once started are not easily stopped, particularly when two governments are involved in the financing, since cancellation could have major political ramifications. In any event, by 1971 it was too late to back off. The first prototype of the Concorde was already being tested, and was being prepared to make a tour of the world's capitals. It was welcomed with energetic hostility in the United States.

The success of American environmentalists in denying the Concorde permission to fly over American territory fanned the political debate. In France some suspected an American plot against the plane. Even more severe a handicap than the prohibition against overflying the United States was the quadrupling in the price of oil that occurred two years before the Concorde was to go into commercial service.

The costs of fuel and maintenance led Air France and British Airways to gradually close their Concorde routes, with the exception of those to New York and Washington—the only ones on which operating costs could be recovered. Five years after the first supersonic route was opened, only seven aircraft out of the fourteen built were in service.[30]

The Concorde will probably go down in history as an example of the mistakes to which excessive technological push can lead. It provides almost a caricature of the technological development pattern underlying many sectors of the sheltered culture—and not only in Europe—in which technical logic outweighs economic considerations. The Concorde being a symbol of technical progress, its backers ignored the practical objections that could be raised against its development. Not until the mid-1970s did they admit that no government could accept the intrusive noise produced by daily breaking of the sound barrier over an inhabited part of its territory. It was also that long before they recognized their mistake with respect to the size of the market for the plane. However, with hindsight, the biggest criticism that can be made of this program is probably that, by draining financial resources away from them, it contributed to the freezing of programs for the development of more traditional aircraft—including the Super Caravelle, which without a doubt would have had substantially better commercial prospects.[31] The Concorde's impact on the rest of aeronautical research remains a controversial subject. It is true that the Concorde program "irrigated" part of civilian aeronautic research. On the other hand, it can be said to have exacerbated one of the natural tendencies within the sheltered culture, and one that is particularly clear in the case of France: the tendency to choose the most sophisticated technical solutions at the expense of economic efficiency. The quest for supersonic performance led to numerous technical feats, but it also oriented research toward expensive solutions that were generally inapplicable to conventional commercial aviation.

European Electronics: Between the State and the Market

Siemens is a large German industrial group that earns two-thirds of its income from electrical and electronics equipment. In 1986 that sum amounted to $23 billion. Turbo-alternators, industrial robots, telephone switching equipment, optical fibers, and medical diagnostic equipment using nuclear magnetic resonance are among Siemens' strong points, and this testifies to the company's competence in high technology. High tech is also prominent in Siemens' other activities. Thanks to its affiliate KWU, the company ranks second in Europe as a manufacturer of nuclear boilers. In association with MBB, Siemens is developing a high-speed magnetically suspended train, which may be operating in Germany at the beginning of the next century. Solidly established in transportation, telecommunications, and electrical equipment, Siemens is particularly representative of the European sheltered culture.[32] With 32,000 researchers out of 363,000 salaried employees, it gives high priority to research. In Munich, where the company has its headquarters, it is said that Siemens has been dominated since its creation by engineers and researchers. These two "corps" have provided most of the group's presidents, including current president Karlheinz Kaske, a former physicist.[33] When Siemens ventures into exposed sectors, it is generally in order to develop products in synergy with its predominant business. For instance, Siemens' involvement with integrated circuits began in an effort to create the electronic components needed for its telecommunications branch. It then diversified to other specialized markets, such as gallium arsenide components used in military applications.[34] For historical reasons, Siemens' presence in data processing has been limited essentially to the mainframe market.[35]

The history of Siemens is characteristic of a certain type of industrial development prevalent among firms in the electronics sector in Europe. Having started behind the shelter of their own borders, and thrived at the outset mostly on public procurement, these firms now must rapidly extend their business internationally because of the narrowness of their national markets. But external expansion is difficult in Europe. European markets are reserved for domestic producers. This is particularly true in telecommunications. The German market is shared by Siemens, Telenorma, and SEL. In Great Britain the market is divided among General Electric, Plessey, and STC. In France it is in the hands of CGE, Thomson, and CGCT. Thus, at the beginning of the 1980s there were no fewer than eight types of digital switching technology being offered on the European market, most of them developed by the various European firms with the support of their respective postal and telecommunications administrations. The European telecommunications industry at the beginning of the 1980s was characterized by extreme protection of national markets controlled by a monopoly or an oligopoly, by the balkanization of industry, and by duplication of development efforts—a situation similar to the one seen ten years earlier in the nuclear sector.[36] Such protection of national markets meant that the competition had to occur outside them. Third World markets were the object of intense competition between Siemens, Ericson, Plessey, Thomson, and CGE. European firms in the sheltered culture thus prepared themselves for battle in the unfamiliar environment of international competition.

Deregulation, which in the early 1980s opened up a large part of the American telecommunications market, displaced the center of gravity in this battle. The United States represents 40 percent of the world market in telecommunications equipment, versus only 15 percent for all of the free markets in the Third World.[37] However, most European manufacturers cannot compete in a market where the entrance fee (that is, the minimal investment required) is very high, as it is in the case of digital switching equipment. In 1985 ITT abandoned efforts to adapt its System 12 for the United States and wrote off about $105 million it had invested in it.[38] The following year CGE made a similar move and closed its plant in Reston, Virginia, which had begun producing small digital switching equipment. Only Siemens and the Swedish firm Ericson persevered. Siemens had been penetrating certain markets in the American sheltered culture, such as those for medical imaging and fiber optics, with remarkable efficiency for some time. In mid-1987, it managed to sell its EWSD switching system to Bell South and Nynex, two large regional phone companies.[39] These contracts represented the Munich firm's first penetration into the heart of the sheltered culture in the United States. Signs of nervousness increased within the American administration with regard to Siemens' activities in the telecommunications field. In late 1986 the chairman of the Federal Communications Commission had written to the managers of the Bell operating companies asking them for information about orders they expected to place with the German firm. For another thing, the Bundespost had become a thorn in the side of the FCC, which had regularly attempted to obtain access to the German market for American telecommunications equipment. Finally, German pressure on the French government to sell CGCT to Siemens rather than to AT&T provoked strong reactions in Washington.[40]

The history of the European data processing industry is a history of missed opportunities. It is also an illustration of contradictory policies of the governments of certain European countries. France and Great Britain considered data processing to be a strategic industry requiring state support and strong protection against foreign competition while at the same time hoping it would serve as a spearhead for exports. These demands quickly become irreconcilable. Computers are not telephone systems. The logic of the sheltered culture, which was so successful for certain European industries in the telecommunications sector, turned out to be disastrous when applied to the exposed culture. The trials and tribulations of the French data processing industry provide an illustration.

The first plan-calcul—a national computer-development project—was launched by the French government in 1965 as a response to a strategic necessity. Bull, the only major French company active in this sector, had just been bought by the American corporation General Electric, which would in turn sell it to Honeywell. In addition, the government of the United States had just refused to supply France with the large computers needed for creating its nuclear deterrent force. The government decided to form CII (Compagnie Internationale pour l'Informatique) with the participation of CGE, Thomson, and Schneider. CII was assigned a twofold mission: to develop the computing capacity needed by the military sector, and to reconquer part of the domestic computer market, which had slipped entirely into American hands.[41] From its creation, CII benefited from a great deal of official attention. Aside from substantial equity financing, the French government gave it $100 million worth of research contracts and loans. Government departments were firmly "invited" to buy their computer equipment from CII. Nevertheless, these attentions were not enough to ensure CII's viability in competition with IBM and Honeywell-Bull, which were solidly established in all the unprotected markets in France. Gradually the idea came to the fore that the French manufacturer should escape from the narrowness of its national market, and that a European alliance would be the way to accomplish this. Thus, in February 1973 the Unidata association came into being. It was supposed to unite the computing capabilities of Siemens, Philips, and CII, and thus to conquer 6 percent of the world market and 20 percent of the European market. But Unidata's first steps were difficult. While the Dutch and Germans were eager to collaborate, the French were more cautious. A certain degree of mistrust existed between French and German industrialists in this sector. CGE, one of CII's stockholders, was frankly hostile to the idea of an alliance with Siemens, its traditional rival in international markets.[42] At the governmental level, some were imputing hegemonic intentions to the German firm within Unidata.

In addition, there was an alternate solution that called for unifying French computing resources, then spread between CII and Honeywell-Bull, via negotiation with the American manufacturer. After much hesitation the Unidata agreement was approved by the French government in February 1974, but 15 months later the American solution prevailed when the political green light was given for a merger between CII and Honeywell-Bull. Unidata was doomed.[43]

Unidata's demise left profound traces in Europe. It reflected the failure of a certain willful conception of technological and industrial development, and it relegated to their respective camps firms that had a clear interest in working together. Paradoxically, the laissez-faire solution and the American alliance that finally prevailed did not mean less government involvement—far from it. Through acquisitions, mergers, and multiple alliances and reorganizations of the French computer industry decreed by successive governments, a solid tradition was established. Industrial policy in the data processing sector came to rest on the principle of reserving public procurement for the "national champion" and its foreign partners. But such a monopoly, inevitable in the initial phase of an industry, contributed to stifling rather than stimulating the innovative and commercial capacities of the champion in question. Government departments paid the cost by being obliged, via succeeding restructurings, to switch from CII equipment to CII-HB, then to Honeywell, and finally to Bull (which again became the French national manufacturer in 1982). The second plan-calcul, which in 1975 merged CII and Honeywell-Bull to form CII-HB, brought the new company $200 million in subsidies and a guarantee of $700 million in public procurement.[44]

Such foulups are not exclusively French. The English and German governments have also tended to use public procurement to support and subsidize a "national champion"—ICL in Great Britain, Siemens in Germany—capable of manufacturing large mainframes, in the purest tradition of the European sheltered culture. Carlo de Benedetti, president of Olivetti (one of the rare European firms to have succeeded in computers using the opposite logic, i.e., that of the exposed culture), is not particularly kind to those responsible for this disaster: "It is probably impossible to make an exact calculation of the sums spent by governments in France, West Germany and the United Kingdom to keep their national mainframe computer companies alive. But the figures certainly run into many billions of dollars, and the end result is well known: today, these three national champions have together less than 5 percent of the world mainframe market, even though they now sell mainly Japanese or American products, while IBM alone has over 70 percent."[45]

This relative weakness of the European computer industry is mirrored in all exposed sectors of the electronics industry in Europe. Among the eight European electronics firms that in 1985 had over $3 billion in revenues, it is significant that only two—Philips and Olivetti—had most of their business in the exposed sector (see figure 7). A similar analysis of Japanese firms gives the opposite result. All nine of the Japanese groups that in 1985 had revenues of more than $3 billion in electronics did most of their business in the exposed culture.[46] It is not surprising that the American situation is intermediate: eleven of the twenty US firms that in 1985 had revenues exceeding $3 billion were predominantly active in the sheltered culture.

Japan's relative specialization in the exposed culture, and Europe's in the sheltered culture, become even more noticeabe when the entire high-technology sector is taken into account. The European aerospace industry largely dominates its Japanese counterpart. The same is true of armaments and, to a lesser degree, pharmaceuticals.

France's Sheltered Culture and Technological Europe

Donald Agger is an old friend of France, and in addition an effictive friend. Since 1969, when he left Lyndon Johnson's administration, in which he had held the post of Deputy Secretary of Transportation, Agger's Washington consulting firm has played an important role in negotiating a number of large Franco-American contracts, particularly in the aeronautics and military fields. For instance, he engineered the close ties between SNECMA and General Electric that led to joint production of the CFM-56. The remarkable commercial success of this aircraft engine (it now powers the McDonnell Douglas DC-8, the Boeing 737, the Air Force's KC-135 and the Airbus A-320) makes it, with nearly 4,000 units sold as of the end

Sheltered culture[1]				Exposed culture[2]		
Electronics sales ($ billion)	Shel-tered	Ex-posed		Electronics sales ($ billion)	Shel-tered	Ex-posed

Europe

1. Siemens	9.70	0.78	0.22	1. Philips	12.4	0.38	0.62
2. Thomson	5.70	0.54	0.46	2. Olivetti	3.5	0.01	0.99
3. GEC	4.50	0.80	0.20				
4. Bosch	3.50	0.79	0.21				
5. CGE	3.35	0.85	0.15				
6. Ericson	3.25	0.77	0.23				

Japan

				1. Matsushita	14.4	0.09	0.91
				2. NEC	9.9	0.37	0.63
				3. Hitachi	9.6	0.21	0.79
				4. Fujitsu	8.3	0.15	0.85
				5. Toshiba	7.1	0.20	0.80
				6. Sony	5.7	0.04	0.96
				7. Sanyo	4.7	0	1.00
				8. Mitsubishi	3.5	0.31	0.69
				9. Sharp	3.1	0	1.00

United States

1. ATT	17.6	0.97	0.03	1. IBM	50.0	0.05	0.95
2. GE-RCA	12.5	0.52	0.48	2. Xerox	9.0	0	1.00
3. Gen. Motors	10.5	0.76	0.24	3. Digital Equ.	7.0	0	1.00
4. ITT	7.0	0.80	0.20	4. H.P	6.6	0.47	0.53
5. Motorola	5.44	0.58	0.42	5. Sperry	5.55	0.43	0.57
6. Honeywell	5.1	0.54	0.46	6. Burroughs	5.0	0.16	0.84
7. Rockwell	3.8	1.00	0.00	7. Texas Instr.	4.35	0.33	0.67
8. Raytheon	3.8	0.95	0.05	8. NCR	3.95	0	1.00
9. TRW	3.3	0.64	0.36	9. Control Data	3.7	0.10	0.90
10. Allied Signal	3.0	0.52	0.48				
11. Westinghouse	3.0	0.87	0.13				

1. Industrial electronics + telecommunications
2. Components + data processing + consumer electronics

Source: Commissariat général du Plan, "Électronique" (December 1986).

Figure 7   Distribution of electronics firms between the two cultures (electronics sales over $3 billion in 1985).

of 1987, the biggest seller of the century for the French aeronautics industry in the United States.[47] Donald Agger also played an important role in the negotiation, in April 1978, of the first contract between Airbus Industrie and Eastern Airlines. He also advised Thomson in 1984 and 1985, when it sold the RITA digital campaign telecommunications system to the US Army. Thus, over the years Donald Agger has come to be a specialist: he is probably one of the Americans who best know the potential of French aeronautical and weapons technologies. He also knows the French government technical elite that most often promotes and sells them in the United States: the armaments engineers. Aside from the fact that they share power over the prestigious Délégation Générale à l'Armement, one of whose prerogatives is to guide French international policy with respect to weapons exports, the armaments engineers also hold a number of key jobs in large French concerns in aeronautics, weapons, and military electronics. During one of the periodic meetings of the armaments engineers, held in Paris in April 1985, Agger could not resist drawing a portrait of the typical French salesman of military and aeronautical equipment: "The armaments engineer . . . first of all is convinced that what is right and just will win, and he is certain that his potential customers subscribe just as religiously as he does to engineering truth . . . If his product is good, he believes it should sell itself. So he will tend, after proposing it, to simply sit back and wait for the inevitable outcome. If success is not immediate, the reason is that the customer is either an idiot or a fool."[48]

This impertinent description does not mean that French firms have not enjoyed brilliant export results in the weapons field. The reason is that technical competence and a capacity for dialogue between similar administrative bodies count for more than the salesman's qualities in this sector of the sheltered culture. French weapons engineers are praised by their Pentagon counterparts. The dialogue between buyers and sellers within the sheltered culture is, in effect, more than anything else a technical exchange between engineers, dealing with specifications and performance. And French engineers are (as is well known in France) the world's best.

French technical competence is rooted in a long tradition to be found in most sectors of France's sheltered culture. In each of those sectors there is a corps of state engineers, the members of which are recruited at the highest level of the educational system—generally upon graduation from the Ecole Polytechnique. They use their influence not only within the administration but also in industry. The great corps of engineers were created long ago as a way for the state to control and master the development of infrastructures and strategic industries. For example, the Corps des ingénieurs des Ponts et Chaussées was created in 1716 to train professionals for government civil engineering and transportation projects. The Corps des Mines was started in 1783 to manage the mining sector; its role was later enlarged to cover all energy-related activities.[49] More recently, the Corps des ingénieurs des Télécommunications was formed to meet the government's need for competent personnel to handle the country's telephone system. The fact that the French government has corps of highly qualified engineers in most fields, recruited among the elite of the nation, is not separable from the fact that the sheltered technical culture is flourishing in the country. This French-style sheltered culture has its roots in a solid Colbertist tradition. In a country where the State is in the habit of regulating everything, it is not surprising that it should have taken charge of technological development. From the canals of the eighteenth century to the nuclear power plants of the twentieth, via the railroads of the nineteenth, the French corps of engineers have always been concerned with making the best use of state-of-the-art techniques to equip the country. A quest for technical performance is in fact one of the characteristics of French-style grand projects.

The numerous "bridges" by which corpsards (as the members of this state corps are called) can move between the public sector and industry are probably a singularity of this system, and more generally, of the French sheltered culture. The firms have an interest in recruiting engineers from the great state corps. The engineers serve to increase the links to the existing administrative structure, provide the firms with knowledge of the government's mysterious ways, and constitute an effective channel for access to information and for the influence needed to win future contracts. For corpsards, the companies generally offer good prospects for a well-paid and professionally interesting career. The most brilliant among them may hope, directly upon leaving the government, to land a job as general manager or even president of an industrial group. The leaders of each corps watch to see that certain important jobs do not leave the corps's zone of influence. Thus, the presidency of Elf-Acquitaine "belongs" to the Corps des Mines, that of Aerospatiale to the Corps de l'Armement, and that of CGE to the Corps des Ponts et Chaussées.[50]

The omni presence of the state, and the state-company symbiosis, are at once sources of strength and weakness of France. Their major merit is to have brought about an integrated model of management that has proved its effectiveness in carrying out certain projects of national interest. The classic example is that of the French nuclear program. Its success results in great part from strong technical expertise within the administration, shared by two powerful public agencies: the EDF, responsible for investments and public procurement, and the CEA, responsible for developing fuel-cycle and safety technologies. Another reason for the success of the nuclear program is the strong coordination at the highest levels of government under the aegis of the Ministry of Industry.[51] Holding key positions at different levels within the nuclear program, engineers from the Corps des Mines handle most of the coordinating functions, up to the highest decision-making level. From 1978 to 1981 an eminent member of this corps, André Giraud, held the post of Minister of Industry. The success of the French nuclear program also derives from the close cooperation that was established between EDF and Framatome, the manufacturer of nuclear boilers—a cooperation that allowed for a degree of standardization in nuclear technology such as is found in no other country. The choice of an engineer from the same Corps des Mines to be the chairman of this industrial group probably had something to do with the efficiency of this relationship.[52]

A similar pattern has been evident in most other major government projects. This all-powerful presence of the state worked to strongly polarize French technological and industrial development. In a country with scattered and often archaic industries, such as France was in the 1960s and the 1970s, this concentration of resources contributed to creating poles of technical excellence, which are today the French sheltered culture's greatest assets in sectors such as armaments, space, aeronautics, nuclear energy, ground transportation, and telecommunications.

This polarization also underlies one of France's weaknesses: because of the small size of the country, the polarization of financial, intellectual, administrative, and industrial resources has been made at the expense of other sectors—particularly sectors of the exposed culture. Even more important, the selection process of the French elite and the logic of its great corps have provided strong incentives for the most capable engineers to work for government rather than industry. When speaking of the relationship between the American defense contractors and the Pentagon, I noted that a firm specializing in public markets tends to mimic the administrative structure it depends on for its contracts. In France, if the osmosis between the government and its industrial contractors has often led to efficiency in the management of large government programs, it has also occasionally contributed to transforming them into appendages of the administrative structure. Within the French sheltered culture there was often little to distinguish between companies' own management rules and those of the government bodies they were dealing with.[53] As a result, the firms in this universe were not well prepared—to say the least—to leave their comfortable nest and face international competition.

There was another major drawback to having financial resources and decision-making power for technological development concentrated in the hands of the government. When it decided to support the growth of exposed industries such as data processing, the government used the logic it was familiar with: that of the sheltered culture. It is small consolation that in Europe this error was not limited to France.

Furthermore, since the rationale for this state-based technological development was national sovereignty, or the need to equip the country, it did not always occur in the sectors providing the best export opportunities for French industry. Although in domestic markets the government was in a position to impose the use of these technologies insofar as government bodies were its end users, the technologies did not necessarily meet with effective demand abroad. The Concorde and nuclear power stand as reminders of this.[54] However, there are important exceptions. It must be emphasized that during the 1970s, several sectors of the French sheltered culture—armaments, telecommunications, urban transportation equipment, electrical machinery, military and civilian aircraft— became within a few years the spearhead of French exports. There are two reasons for this: an absolute need to survive for certain firms, but also a true comparative advantage of these firms in attempting to penetrate the sheltered culture in foreign countries.

In the late 1960s—unlike their American counterparts, which could expand in a very large sheltered domestic market—companies in the European sheltered culture already needed to escape the confines of their national markets and sell their products abroad. Between 1968 and 1972, French manufacturers of military aircraft began to export seriously at a time when their American counterparts were busy filling the Pentagon's needs for the Vietnam war. Let us also remember that at this time other European countries' markets were largely inaccessible to French industry, and that industrial alliances with companies in neighboring countries were discouraged by governments because of the strategic nature of many sheltered industries. This situation explains the stringent competition among firms in the European sheltered culture for the conquest of Third World markets, the only ones that seemed easily accessible.[55]

The combined talents developed within the French sheltered culture were effective in this battle. Sales of equipment such as nuclear plants, telecommunications networks, and civilian or military aircraft are usually negotiated between governments, sometimes falling within bilateral protocols for cooperation. Such negotiation is the only access to the market, and it provides the political and financial framework required for the companies' action. The reasoning that often prevails in this context consists in setting up a package of equipment, technologies, and services that fulfill the customer's needs, and then attempting to build a financing arrangement. What is most important is to satisfy the mandatory conditions for making the sale. The cost, in these circumstances, is generally of secondary importance. If the selling price of a particular piece of equipment turns out to be lower than its actual cost; however, this is not necessarily an obstacle to the transaction, since the government of the selling country may subsidize the difference. For the government, what counts is not the profit for one particular company but rather the potential impact of this government-to-government transaction on turnover in the industry as a whole. This is a familiar approach for industries in the sheltered culture. It enables them to increase revenues, and it provides economies of scale associated with larger series—and they may still obtain compensation from the state if forced to sell at a loss.

France, of course, is not the only country to apply this sort of "coordination" of exports to developing countries. Most industrialized nations have been enthusiastic practitioners of it.[56]

This approach needed only slight adaptation when companies of the European sheltered culture tried to gain a foothold in American sheltered markets. Government negotiation was in some cases replaced by direct dialogue between companies and their US customers, but the selling state was often there behind the scenes, providing export subsidies and long-term, low-interest loans to compensate for the temporary operating losses that are inevitable when a company attempts to penetrate foreign sheltered-culture markets.

The company's attitude resulting from government backing is just as important as the financial dimension. There is a considerable advantage involved in being able to attack a new market without having to worry about immediately making adjustments between sales price and costs. One can be commercially aggressive with a policy aimed only at long-term financial equilibrium—a luxury that an industrialist, whose survival may depend on each contract, cannot allow himself. Paradoxically, the European sheltered companies were well prepared to use price competition in American sheltered high-technology markets.

In some sectors of the European sheltered culture, such policies were nevertheless not enough to ensure the survival of national companies. In spite of the disappointments it led to—Euratom, Concorde, Unidata—European industrial cooperation was seen as the only means of pulling together the resources needed for ever-higher production investments and for sharing the growing costs of research and development.

Involved as it was in expensive technological programs that could hardly be justified within the narrow framework of its national market, France had more interest in cooperation than any other country. But such cooperation was impossible as long as each of the potential partners intended to use it to the detriment of its neighbors—a situation that has prevailed too long in Europe (for example, in the cases of nuclear power and telecommunications).

Aeronautics is the first sector in which these national visions have been overridden, thanks to the Airbus program.[57]

The European aeronautics consortium was established in 1970. Its future was threatened before its birth. At the end of 1968, Great Britain, whose contribution to the A-300 (the first airplane built by Airbus Industrie) was to be the engines, decided to withdraw. Rolls-Royce preferred to manufacture engines for Lockheed's L-1011.[58] The project was saved when General Electric became the supplier of engines. But the main reason that Airbus survived this crisis is that it was built around solid Franco-German cooperation. The two countries had already joined forces in the 1960s to produce the Transall military plane. In the aeronautics field they had common ambitions and complementary advantages: France had the technological and industrial expertise; Germany had financial resources.[59] Since World War II the aeronautics sector had occupied only a modest place in the German economy, as a result of the prohibition forbidding Germany to develop any defense-related industrial activities. It was now time to make up for that. Cooperation with France would provide the occasion. In an increasingly capitalistic sector, this was a means of creating critical mass, making it possible to recover development costs and production investments. It was also a way for French and German companies to expand beyond the narrow framework of their national markets—an indispensable step toward reaching the world market, 95 percent of which was dominated by American manufacturers. France remembered its experience with the Caravelle; although it was one of the most innovative planes of its time, the opportunity to capitalize on its initial success (280 units sold in numerous countries) was lost because of a lack of resources for developing new generations of aircraft. Most available resources were at that time mobilized for the Concorde program.

From the start Airbus Industrie overthrew the logic then still in effect in the European aeronautics industry. Instead of designing the most advanced airplane, as the Caravelle and the Concorde had been in their time, the company at the outset identified a need on the world market: that for a medium-range twin-engine airplane with large capacity. This strategy paid off. Sales of the Airbus A-300 progressed rapidly, with 176 orders by the end of 1978.

This success was due to several factors. First of all, the A-300 was a good product. It was an excellent compromise between tried and tested technical solutions and novel ones. Second, the use of two engines—a major innovation in the area of jumbo jets—led to lower operating costs. Airbus thus managed to occupy a niche, ensuring temporary protection against competitors, who could offer only smaller aircraft for 150 passengers or more expensive ones for 300. The Franco-German venture had not only made a technical breakthrough but had also produced a commercially efficient product. A third factor was the result of an aggressive marketing and price policy in an industry that was not at all accustomed to such an approach. The fourth factor, inherent in the American market, was fortuitous: in 1978 Eastern Airlines was on the brink of bankruptcy, and Airbus' financially seductive offer provided the airline with an opportunity to renew part of its aging fleet.[60] A last (but important) factor is related to the restructuring of world demand that occurred during the 1970s. A new generation of airlines was emerging, principally in the Far East (Korean Airlines, Thai Airways, Japan Airlines), and these new airlines were striving to catch a share of the Western airlines' market. They needed equipment, and the Airbus was well suited to their needs. Their expanding business contributed to displacing the center of gravity of world demand, which until then had been located in the American market. Taking advantage of their enormous domestic market, American aircraft manufacturers had made little effort to promote their products in distant markets. Their dominant position worldwide had been acquired without any particular effort to export.

The European governments' policy of supporting Airbus was, of course, also a factor in this success. While still in the cradle, Airbus Industrie received the means to develop the A-300. Long-term loans, repayable in the event of success, enabled Airbus to meet its growing financial needs for the development of new generations of aircraft. The organization of Airbus Industrie was that of a Groupement d'Interêt Economique, a French legal arrangement by which each government can directly support its own companies participating in the joint program. This support reflects the various firms' levels of participation in the consortium: 37.9 percent for the French Aerospatiale, 37.9 percent for Deutsch Airbus, 20 percent for British Aerospace, and 4.2 percent for the Spanish firm CASA.[61]

Government involvement in the Airbus program has not been limited to the financing of technological development; it has also included low-interest loans and export subsidies, which proved to be critical during the first years of Airbus Industrie's commercial operations.

In civilian aeronautics, the decisive buying criterion is less the aircraft's basic price—the "sticker price," which all manufacturers adjust to meet their competitors' prices—than the financing package being offered the airline or the buyer country.[62] We can see how it was that, in using commercial practices from the exposed sector in the sheltered US aeronautics market, while continuing to benefit from the European governments' financial support, Airbus Industrie aroused complaints from firms that had until then been dominating these markets. Seen from Washington or Seattle, the impudence of Airbus Industrie lay not in its having succeeded in building airplanes as efficient as those of Boeing or McDonnell Douglas or in its having persuaded a few North American companies to buy them but rather in its having transgressed the implicit rules of a key sector of the sheltered culture, which had until then been practically a worldwide American monopoly. Seen from Bonn or Paris, American complaints about unfair competition because of "loans" in the form of subsidies to Airbus Industrie seem slightly hypocritical. Over the past twenty years, where would McDonnell Douglas, whose civilian branch was able to generate a profit only three times, have been without Pentagon procurement? The Air Force's purchase of sixty KC-10 military transport planes prevented the almost inevitable closing of the production line for the DC-10, the KC-10's civilian counterpart. It not only enabled McDonnell Douglas to remain in the civilian market for jumbo jet but also provided some of the cash needed to finance the development of the MD-11, the next-generation civilian aircraft. A similar scenario will occur with respect to the C-17, another military transport, 200 units of which will be built in the 90s. The General Manager of McDonnell Douglas himself underlines the importance of this new military program to the future of his company's civilian branch: "Recent Congresssional approval for financing the C-17, a military jet transport to be built by McDonnell Douglas, reduces the enormous risk involved in developing a large commercial jet. Not only will the C-17 program maintain our aeronautics activity in Long

Beach, but by increasing our cash flow the C-17 program will enable us more easily to absorb potential losses from the MD-11 program."[63]

The second sector in which France was able to draw its partners countries into a joint effort is space. The Ariane program, launched in 1973, was a response to political and strategic concerns that Europe should have its own capability to orbit satellites. France's point was that, in the era of telecommunications satellites, Europe could not continue to be completely dependent on the United States for launch vehicles. This proposal did not immediately win over all the other European countries, but gradually a good number of them joined the program—all the more willingly since it was becoming clear at the end of the 1970s that Ariane could be targered toward the international market for commercial launch services.

Using the lessons learned from the failure of Europa, the earlier attempt to create a European launch vehicle, the Ariane project was set up as a carefully integrated project under the auspices of a single executive organization, the Centre National d'Etudes Spatiales. The French space agency was at that time the only organization in Europe capable of assuming responsability as prime contractor. The European Space Agency, created in 1975, had the responsability for getting financial participation from the thirteen countries that joined the program. The French contribution was the highest (53 percent), followed by that of Germany (20 percent). British participation amounted to only 3.5 percent.[64]

Ariane's first trial took place in December 1979. The launch in October 1983 of a telecommunications satellite for Intelsat marked its operational debut. By the end of 1986, 63 launches, worth more than $2.5 billion, had been booked.

Part of Arianespace's success stems from its original organization, providing unified command and integrating industrial participation from a dozen countries. Other factors, as we have seen, include the fact that the European booster filled a technological niche. Developed from a set of judicious technical options, Ariane was a timely response to a specific need. In addition, Arianespace was able to take advantage of the disastrous choices made by its main competitor, NASA. Another factor behind its success is purely political. Even though France was forced to make an unrelenting

Table 4
Distribution of the public R&D effort between high-technology industries and traditional industries in various OECD countries in 1980. Source: OECD science and technology indicators, 1986.

	High-technology industries	Traditional industries
United States	88%	12%
France	91%	9%
Great Britain	95%	5%
Germany	67%	33%
Japan	21%	79%

effort to convince its partners, Ariane is still the result of a common political goal among Europeans. Little by little this program has become the symbol of a Europe at long last capable of successfully joining forces in the area of advanced technology.

It is no accident that this result was obtained within the sheltered culture. Once agreement has been reached on a common goal, it is in principle easier for countries to cooperate when their governments are fully mastering the game. Although space programs are largely dependent on public financing, their high cost forbids strictly national programs. In the European space industry, as in aeronautics, cooperation was the only answer.

Heritage and Strategy

Europe does best in the sheltered culture. Airbus and Ariane illustrate this point. This Old World specialization with respect to the two cultures should not be surprising in view of the analysis of high-technology markets in earlier chapters. In Europe as in the case of Japan, such specialization is the result of comparative advantages inherited from history, and of political will.

The postwar priority given in France and Great Britain to building a nuclear deterrent force led to technological development modeled after the American sheltered culture. Government priorities for military and civilian technological development in these two countries were, in the 1960s, similar to those in Washington, as table 4 illustrates. They contributed to the creation of technical capabilities in different sectors of the European sheltered culture that were tied directly or indirectly to the defense effort: civilian and military aeronautics, armaments, civilian nuclear power, space. But unlike the United States, where there was a more broadly self-financed industrial R&D effort, France and Britain made progress in those industries at the expense of technological development in others.

The European R&D effort in the sheltered culture shows two other fundamental differences with its American counterpart.

The first is the more limited availability of resources. At the end of the 1960s the total volume of resources devoted to R&D by European Economic Community amounted to half the American effort.[65] These financial limitations were aggravated by the duplication of both civilian and military technology-development efforts in European countries. With the exception of the Concorde program, the French and the British followed parallel routes, and most of the time they were competitors. Governmental technological choices, therefore, had by necessity to be much more selective than in the United States. Most of the time the available resources did not even allow for studying alternate technical solutions, as they did in America.[66] Large technological programs in Europe often led to high-risk technical gambles. As a result, governments often embarked on uncertain undertakings and became prisoners of the technical choices from which they derived. In these circumstances it is always difficult to stop a project, even if a chosen option turns out to be less promising that expected. The supersonic aircraft of the late 1960s and the breeder reactor of the 1980s are two characteristic examples of such a situation.[67]

The second difference between Europe and the United States is to be found in the weakness and the dispersion of European high-tech industries. As a consequence, while government programs in the United States mobilized only a limited fraction of industry, in Europe they absorbed most of the available industrial resources, hence contributing strongly to polarizing European technological and industrial capabilities.

The development of the sheltered culture in Europe is also the result of a solid scientific tradition. It finds expression in the large public research laboratories, whether of fundamental or applied character, that are an outstanding aspect of the European research system. More specific factors have also favored the development of the sheltered culture in certain countries. In France, the strong tradition of government involvement in the country's economic life, the centralization of government decisions, and the existence of a highly competent technical elite and of powerful applied research agencies within the government have also favored the development of the sheltered culture. Some of these factors are also to be found in Great Britain, although there are notable differences in the way technological programs are managed there.[68]

The situation in Germany is very different. Busy reconstructing its industrial and economic infrastructure after the World War II, Germany did not launch a vigorous R&D effort until the early 1960s. As in France and Great Britain, a significant part of government R&D resources was devoted to the development of large infrastructures, in which the sheltered culture flourished. In 1970, 65 percent of public financing was devoted to energy, transportation, space, and military research.[69] Civilian nuclear energy was given particular priority with $450 million devoted to this sector in 1975 (30 percent more than in France).[70] Although immediately after the war Germany was forbidden to develop technologies tied to defense, its government did not wish to preclude such developments when they had civilian applications. Siemens is a model for firms in the German sheltered culture that served as industrial partners for these government priorities. However, alongside these developments, a large part of public R&D resources—33 percent in 1980—was used to promote innovation in other sectors in which German has traditionally shown great competence: the chemical, mechanical, machine tool, and electrotechnical sectors. Decentralization of many public decisions at the level of the Länder (states) and a long tradition of ties between industry and the universities also favored this effort.[71]

Corresponding to these priorities in technological policy, which are significantly different from those in France and Great Britain, Germany has a different industrial fabric. Alongside the very large industrial groups—all of which, in the technological sectors, belong to the sheltered culture—there are a number of medium-size companies, such as Nixdorf in electronics, that are characteristic of the exposed culture in Germany. In fact, if one looks a little deeper into German industry, one discovers a mixed situation, with companies in each of the two cultures well represented. The German exposed culture is also vigourous in more traditional industrial sectors, such as automobiles. More generally speaking, there is a solid entrepreneurial tradition in Germany that provides an important advantage for the development of an exposed industrial sector. In addition, aside from certain fields (such as energy, transportation, and telecommunications) in which public procurement plays a determining role, the tutelage and influence exercised by the government over industry are not comparable to what they are in France or Great Britain. However, all this was not enough to allow for the development in Germany of world-class industrial groups in the exposed sector. It is probably not by chance that the only large European electronics firms to have emerged in exposed sectors—Philips and Olivetti—developed in two countries—the Netherlands and Italy—in which there was a solid entrepreneurial and mercantile tradition, and where the government's strategic choices did not have any impact on those made by the industrialists. Forced from the outset to enter the world market, these two firms probably had no choice but to adapt to the logic of the exposed culture or disappear.

The extensive development of both cultures in Germany says a great deal about that country's ambivalence with respect to industrial and technological Europe. Germany's exposed sector emphasizes pure free-market logic, which does not necessarily suit the objectives of European construction. Its sheltered sector praises common objectives but still must accept tradeoffs involving longstanding antagonisms among many of its industrialists and their neighbors and their obvious convergence of interests at political or economic levels in the EEC. The chief merit of the French proposals that led to the Airbus and Ariane programs was that they made use of these converging interests. Aside from the objectives that the governments shared, what gave strenght to these programs was the companies' interest in building industrial alliances. This has happened too seldom in European high-tech industries, particularly in the exposed culture. In consumer electronics or data processing, for example, rivalries among European firms have been almost endemic for the past twenty years, with each of the "national champions"—

French, British, or German—finding it easier to seek industrial partners in the United States or in Japan. Thanks to Airbus and Ariane, France has, in a certain sense, contributed to European specialization in the sheltered culture; however, this was probably a necessary first step toward creating a technological Europe.

Irresistible Domination

10

As a designer of large computers for Univac and Control Data, Seymour Cray was already a celebrity when he decided to found Cray Research in 1972. Like many others, he started his own business because his ideas were going unnoticed in the firm that employed him. For several years he had been thinking about new computer architecture that would provide a giant leap forward in the performance of large scientific computers. For Seymour Cray the future was to be found in vectorial machines, built so as to handle multiple operations on data organized as vectors. In this way the speed of calculations could be considerably increased, perhaps up to several megaflops (in other words, several million arithmetic operations per second). When Cray discussed his supercomputer project with his boss, William Norris, president of Control Data, he was listened to with interest. Norris was also a lover of computers. However, he quickly raised a major objection to Cray's project: the market. Aside from a few laboratories working for the Pentagon, who could have computation needs to justify such investments? In spite of that, Seymour Cray remained convinced that he was right. He decided to leave Control Data and set up his own company. After making the rounds of a number of large research laboratories, he was in possession not only of the specifications for his future machine but of his first orders as well. In September 1976 the first Cray-One, the prototype of a series of supercomputers, was delivered to Lawrence Livermore Laboratory. The second was financed in 1977 by the National Science Foundation for the

National Atmospheric Research Center in Boulder, Colorado.[1] So there were others besides the military who needed "number crunchers," as supercomputers came to be called, even though most of the orders for the Cray-One did come from large laboratories working for the Department of Defense.

In the early 1980s the large universities also became interested in supercomputers. They made it possible, on a scale never before seen, to develop techniques of numeric simulation making it possible to avoid costly experiments in many research sectors. Control Data finally became interested in this technological niche, which gradually became a high-growth market. In 1980 it brought out the Cyber-205, with a maximum performance—better than that of the Cray-One—of 405 megaflops. In the meantime, Seymour Cray, with the help of a brilliant computer specialist of Taiwanese extraction named Chen, whom he had hired in 1979, brought out his new Cray-Two series, with a calculating power of 1,800 megaflops. A race for more powerful machines had effectively begun.[2] It was made possible by the development of multiple processors and ever-faster components. In spite of their high price, the demand for Cray machines had by now spread to a large number of engineering firms. (The Cray-Two was priced between $15 million and $20 million.) Using the fine logic of the sheltered culture, Cray maintained his technological lead, and his captive clientele remained with him because of the superior performance of his equipment and his excellent after-sales service. However, by expanding, the supercomputer market had become attractive to other computer manufacturers. In 1984 the Japanese companies Fujitsu, NEC, and Hitachi added supercomputers to their catalogs. Their sales did not yet extend beyond Japanese territory, a market that Cray Research found it very difficult to penetrate. But in 1986 there came news that shook the profession: NEC had just sold its first supercomputer on American soil, to the Houston Area Research Center of the University of Texas. As usual, the Japanese firms were offering equipment with lower performance but at an unbeatable price.[3] They were attempting to introduce price competition for the first time in the heart of the sheltered sector of the American computer industry.

At the end of 1986 Cray was in an enviable position. The firm had sold 154 of the 244 supercomputers in use around the world.

It employed 4,000 persons. It had generated profits of $125 million on sales of $597 million. It showed the best rate of profit in the entire American computer industry.[4] But Cray Research was going through a serious crisis that climaxed with Chen's departure in August 1987. Chen took with him the idea of a "super supercomputer," a machine that would use gallium arsenide components and internal links of superconducting materials, which would make it possible to increase the transfer speeds even further.[5] The CHEN-MP—as the machine, scheduled to come out in 1992, was to be called—would reach the phenomenal speed of 35,000 megaflops, 250 times faster than the first Cray.

Cray Research today is at a crossroads, astride two cultures. By staying in the race for technological performance, the company can certainly maintain its dominant position for some time, but it will leave an increasing share of the market to firms that have chosen a price-performance compromise. The arrival in this field of new competitors such as IBM, Amdahl, and Sperry, and the resurgence of Control Data thanks to its supercomputer affiliate ETA-Systems, shows that the market structure is changing. Even though one can still be skeptical for the time being about Japanese commercial penetration in this sector, price competition will probably develop under the impetus of the newcomers, almost all of whom come from the exposed sector. The departure of Chen, an unconditional proponent of higher-performance machines, although costly for Cray Research, does not mean the company has decided to leave the comfort of the sheltered culture. But it does show that questions are probably being asked at the top of the Minneapolis firm about which strategy to use.

This transformation of the supercomputer market is characteristic of changes now affecting many sectors of the American sheltered culture. It is happening under pressure from both Japanese and European firms. The Japanese proceed, as we have seen, by progressively enlarging the field of the exposed culture. Able to design increasingly complex products of increasing technicality, Japanese companies use with these new products the same approach that enabled them to succeed in their first forays into the American exposed culture: lower production costs, rigorous quality control, and aggressive marketing based on unbeatably low prices.

Table 5
Market shares of foreign firms in the US sheltered
culture.

Market	Foreign firms	Share of US market (%) (1987)
Launchers	Arianespace	50
Public switching equipment	Northern, Siemens, Ericson	25
Telephone transmission	Siemens, Alcatel	10
Medical imaging	Siemens, NEC	10
Civilian aeronautics	Airbus Industrie	15
Aircraft engines	SNECMA, Rolls-Royce	25
Pharmaceuticals	Ciba-Geigy, Bayer, Hofmann Laroche	6
Military electronics	Thomson-Plessey	1
Armaments	Rolls-Royce, Beretta	0.5

Sources: French Trade Commission, French embassy, Washington, DC, and
miscellaneous others.

European firms have also adopted a similar approach in sectors
of the sheltered culture such as aeronautics, public telephone equipment, and military electronics. Table 5 shows the principal foreign
firms that do business or have an affiliate in the United States and
that are present in the sheltered culture's major markets. It also
shows their market share as determined by information available in
late 1987.[6]

Step by step, the pressure of international competition has
imposed the rules of the exposed culture on more and more sectors
of the American high-technology sheltered culture. Two other
important factors have also contributed to extending the field of
the exposed culture: technological evolution and deregulation
policies.

Technical Progress and the Expansion of the Exposed Culture

Whenever demand is elastic with respect to price, the rationale of
industrial managers is to manufacture and sell large amounts of
products by reducing production costs. The history of industrial

development shows that this objective was reached in a large number of cases through innovations that transformed the manufacturing process. The classic example is Henry Ford's first assembly lines, which considerably lowered the selling prices of automobiles. In other cases, access to a larger number of users was possible because of radical changes in some characteristics of an existing product. This is what has happened with computers. Two innovations that helped usher in the era of mass distribution of computers were the invention of the microcomputer (which, because of its low cost and ease of use, considerably enlarged the field for computer applications) and the progressive adoption of languages and interfaces that made computers accessible to nonspecialists. Initially designed as a complex product that could be used only by a highly trained elite, the computer became simpler and more everyday. The logic of the sheltered culture, which persisted for some top-of-the-line products in the computer industry, has tended to recede as the relative size of the mainframe market has fallen, to the advantage of mini- and microcomputers (see figure 8).

A similar evolution is occurring in many segments of the telecommunications market. Public telephone equipment (for switching and transmission), which makes up the sheltered part of this sector, now represents no more than half the total market in the United States, as opposed to two-thirds ten years ago.[7] This change has come about as a result of the explosion of the private telephone market and the progressive integration of telecommunications and data-processing terminals. Telecommunications systems have become decentralized with the multiplication of private and local networks and the diversification of equipment. The performance and service that can be provided by small telephone exchanges (PBX, Key Systems) have increased, while their price has fallen considerably. Telephone answering machines, telecopiers, and other consumer-premise equipment, the production of which obeys the logic of the exposed culture, have increased in number and variety. The market for them is expanding as competitive pressure forces their price to fall. The many innovations combining data processing and telecommunications have also contributed to this diversification of products and services available to both individuals and corporations.

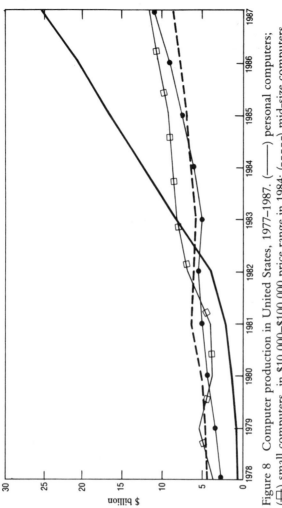

Figure 8 Computer production in United States, 1977–1987. (———) personal computers; (⊞) small computers, in $10,000–$100,000 price range in 1984; (- - - -) mid-size computers, in $100,000–$1,000,000 price range; (●) large computers, over $1,000,000, including super-computers. Sources: IDC, 1984, OCDE, 1985.

In the field of data-processing services I have also underlined the irresistible progress of mass-produced software packages at the expense of custom-made programs. This evolution is irreversible, even if the latter are still dominant in the software business.

In the pharmaceutical industry, the development of generic drugs has also favored extension of the exposed culture. These are products that are no longer covered by patents, which provide protection for only 17 to 20 years in the United States. They can therefore be produced by a larger number of firms. The procedure for obtaining marketing approval in this case, is highly simplified. The manufacturer need only prove that his product is identical to the one that was authorized earlier. Intense price competition is arising in this rapidly growing market, which increased from $500 million in 1983 to $1.3 billion in 1986.[8] The firms emerging in this market are not the large pharmaceutical companies of the sheltered sector. Mary Beals, an analyst with the New York firm Drexel Burnham Lambert, notes that "most of the major brands cannot compete with the manufacturers of generic drugs because of their pricing policies."[9] This observation should not surprise us. The generic drug market is dominated by another logic—one that is coming ever closer to the logic of the exposed culture.

New Rules of the Game in Telecommunications

Judge Harold Green is very interested in French telecommunications technology. Some of his latest decisions were based on the French example, for it was with reference to Minitel that, in October 1987, he authorized the Bell operating companies to experiment with videotex services—a decision they had awaited impatiently, since they wanted to diversify their activities in added-value services.[10] Judge Green is a short, affable man in his sixties. He speaks readily, when the subject is brought up, about the case that lifted him out of anonymity at the Justice Department. It was he who (with his soul and conscience as guide, he says) made one of the most important decisions in the industrial history of the United States: the one that led to the break up of AT&T. In order to understand a situation that in European eyes seems truly surrealistic—since it enabled "one little judge" to dismember the world's most powerful industrial empire—a brief look into the past is mandatory.

During the mid-1960s, some 80 years after its birth, the Bell System was at the pinnacle of its power. The telephone industry's growth rate was higher than that of the GNP. Technical progress and economies of scale had made it possible to regularly lower the cost of long-distance communications. The cost of a one-minute call between New York and San Francisco, $2.50 in 1950, had fallen to $2.00 in 1965—a 65 percent reduction in constant dollars.[11] One out of every two Americans had a phone. Telephone service had become almost universal, and the United States had the world's best service.[12]

AT&T's profits had never been higher, but its status as a quasimonopoly was coming under increasing attack. Over the years the company found it more and more difficult to forestall the many initiatives aimed at challenging its position by means of new ideas, products, and services for telephone users. In the 1950s, AT&T had successfully opposed the firms that had tried to market the first answering machines. The FCC upheld their interdiction, siding with AT&T's claim that their use could damage the network. This was the key argument that the company used against all outside innovations. In 1968, however, it lost its battle against the Carter-phone, a simple device for connecting a radio to a telephone. The Carterphone made it possible for any owner of a radio transceiver—and there were many of them on ranches and in isolated parts of the United States—to hold a conversation with any phone sub-scriber. By allowing Thomas Carter to market his Carterphone, the FCC created an opening in the closed territory controlled by the Bell System. Five years later, after a series of legal battles, the phone company was forced to accept the user's freedom to connect the equipment of his choice to the network, provided it contained an approved protective circuit.[13] Furthermore, in 1973 the FCC made a formal distinction between services that would be subject to regulations and unregulated services (including, in particular, the connection of data-processing equipment). While independent phone company subsidiaries were authorized to market these new services, AT&T was prohibited from doing so.

The second major breach in AT&T's quasi-monopoly came with the authorization in 1971 of free competition among Specialized Common Carriers (SCCs)—companies wishing to make commercial use of private point-to-point microwave telephone links that were independent of the public telephone system. This service was of interest to certain large companies with heavy telephone traffic between headquarters and subsidiaries or plants. Microwave transmission could be offered more cheaply than the Bell System's cable network because of the tariff regulations to which the latter was subject. However, the firms offering this new service—MCI and Sprint—soon discovered that they would be unable to recover their investments if they limited themselves solely to point-to-point phone service. For this reason they created subsidiaries able to switch any phone call from the local public network to their own long-distance links. At the other end of the line, calls could be switched back to the public network to reach their destination. AT&T protested energetically and accused the new companies of unfair practices: in effect, not limited by regulations, they were offering their services on the most profitable links, whereas the Bell System was constrained by public-service obligations. AT&T began a long legal battle that ended in defeat in 1978 before the Court of Appeals of the District of Columbia.[14] MCI, Sprint, and the other SCCs were authorized to compete against the Bell System for long-distance telephone traffic.

In the meantime, a Sword of Damocles had appeared over the head of the telecommunications giant. As a result of numerous complaints about abuses stemming from its dominant position, pressed by both dissatisfied users and competitors, the Justice Department in 1974 began an antitrust suit against AT&T.[15] In fact, the Bell System was coming under attack from several sides at once. In 1976, proposals for "anti-Bell" legislation began to appear in Congress. Meanwhile, the FCC was trying, with increasing difficulty, to maintain coherence within a hybrid system that was the by-product of contradictory demands from the federal government. Within this system were to be found both the regulated quasi-monopoly of the Bell System (which was based on an equalized tariff system still favored by the FCC) and a competitive, unregulated sector (which the government wished to foster as new technologies and their associated services were developed). When

Charles L. Brown took over as the head of AT&T in 1979, he realized that his margin for maneuvering was extremely small. While his company was tied up in administrative and legal tangles, every passing month worked to the advantage of his competitors, who were taking every advantage of their position on unregulated markets. By concentrating their efforts on the most profitable segments in long-distance communications, they were managing to provide service as good as the Bell System's. They were also taking advantage of the restrictions placed on AT&T's business while reinforcing and diversifying their positions in the market for new services, and becoming able to derive profit from the synergy between computers and communications.[16] Nor was AT&T's new general manager particularly optimistic about a possible legal settlement. Judge Green was a tenacious man. There was a chance that the company might lose the suit. Of course, if the problem dragged on, a solution might be provided by Congress in the form of a telecommunications bill. But it might not be any more favorable. Deregulation and competition were in fashion on the Hill. In his book on the American telecommunications industry, Leonard Hyman summarizes the dilemma facing AT&T management: "The Department of Justice did not want the Bell System. Congress did not want it. Why fight for universal service, and be eaten away by competitors in return? Why not go out and make money the way other businesses do?"[17] This is why Charles Brown decided that the best way out would be to negotiate with the government. On January 2, 1982, he called Assistant Attorney General William Baxter and agreed to a compromise proposal that Baxter had sent him ten days earlier.[18]

Baxter, a professor of antitrust law at Stanford University, had been appointed to his post by President Reagan in February 1981. In April, during the course of the antitrust investigation being held under Judge Green, he let AT&T's management know what he wanted. Dismantling, he felt, was inevitable. The minimum condition for a negotiated settlement, he said, would be to separate the regional telephone companies from the Bell System. Given his professed principles, Baxter felt a profound aversion to government regulation of economic activity, as did his economist colleagues of the Chicago School. Since Ronald Reagan's rise to power, they had been exerting a powerful influence over the government. For Baxter, the breakup of the Bell System would have the advantage of limiting government regulation to regional operating companies, leaving the road open for competition in long-distance services and other areas that were still regulated under the AT&T monopoly.

Baxter's proposal was received with astonishment at AT&T, where no one had seriously considered the possibility that the company might be dismantled. The company's lawyers were even hopeful of winning their case.

Judge Green's investigation and hearings continued. This antitrust suit, now in its seventh year, had produced a mountain of documents and testimony on both sides. In all it represented several million pieces of paper, filling three floors of a building that AT&T had leased for this purpose near the Justice Department. Every evening the lawyers for both sides would decide which documents should be presented to the judge the next day, following an established agenda. The documents were organized and photocopied that night, then loaded onto carts to be wheeled to the courtroom. No fewer than 300 employees were needed for this daily task.[19]

Judge Green wanted all the essential documents to pass through his hands. He wanted to see and hear everything. He wanted to go to the heart of the matter. He fully mastered the proceedings, and intended to carry them through diligently. They ended in July 1981, after a marathon of depositions and testimony. AT&T's lawyers decided to move for dismissal. They submitted a 500-page brief in which they requested that all accusations brought against the company by the government be dismissed, and that the case be closed. From their point of view it was clear that the plaintiffs—the government representatives who had accused AT&T of violating the Sherman Antitrust Act—had not made a case against their client. But Judge Green didn't see it that way. On September 11 he gave his verdict: "The motion to dismiss is denied. The testimony and documentary evidence adduced by the government demonstrate that the Bell System has violated the antitrust laws in a number of ways over a lengthy period of time. The burden is on the defendant to refute the factual showings."[20]

Even if it was only a stage in the investigation, this decision was devastating for AT&T. One of the company's lawyers said: "Before AT&T even put in any evidence, we knew we were confronted with a judge who was not hearing our side of the case."

AT&T was in danger of losing its case. For Charles Brown this event substantially changed the situation. His company had already lost too much money in this affair: some $375 million in seven years for a trial to which some 3,000 persons were assigned. Moreover, the Justice Department's example had been followed. A number of AT&T's competitors also filed suit against the company for violating the antitrust laws. During recent years these suits had been piling up at the rate of one every three or four months. And even in the improbable event that AT&T won the case, the firm would at best only find itself back at the starting point, still a prisoner of the 1956 consent decree that prohibited it from entering new markets in the borderline areas between computers and telecommunications. Trading the regional operating companies for the right to enter these new markets would not only relieve AT&T of the burden of the trial but could also better prepare it for the future. This is why, in spite of the seeming intransigence of Baxter's proposal, Charles Brown decided to accept it.

The settlement of the AT&T case was announced on January 8, 1982. Its terms stunned most observers. As Leonard Hyman noted, "The agreement was startling for its severity and for its dogmatic view of the world."[21]

The Bell System was divided into eight entities. AT&T was designated to handle long-distance services, keeping Western Electric and most of Bell Labs and retaining the right to develop new technologies for merging computers and communications.[22] Seven independent regional companies, operating with the status of regulated monopolies, were given responsibility for providing local telephone service. Once the shock had passed, most specialists agreed that Charles Brown had come out remarkably well. In letting go of the regional companies, he had managed to divest his firm of most of the problems that had plagued the Bell System. Certain analysts suspected AT&T of having pocketed "a third of the assets and two-thirds of the profits" that had existed before the dismantling.[23] However, for Charles Brown the most important work remained to be done. His objective was to make AT&T the world's leading information-age company, the first to globally integrate the opportunities provided by the wedding of computers and telecommunications.

Nearly six years after Judge Green's decision, and four years after it became effective on January 1, 1984, the American telecommunications scene was somewhat different from what had been foreseen a few years earlier by the government authorities responsible for deregulation.

The opening of the long-distance market to competition, which had been one of the major objectives of deregulation, has been achieved only in part. In spite of the FCC's efforts to give young companies the same access to users as AT&T (by asking all subscribers to choose their company, after a long and costly advertising campaign in 1986), AT&T had reconstituted its quasi-monopoly within a year. Its share of the long-distance traffic, 91 percent in 1983, was still close to 80 percent in mid-1987. Most of the competing companies, such as MCI and Sprint, had needed to borrow massively in order to build networks capable of competing with AT&T's and were facing serious financial problems.[24]

In the sector of large switching equipment for the regional Bell operating companies, AT&T's market share decreased in 1985 to the benefit of its major competitor, Canadian Northern Telecom; however, AT&T regained part of the lost business and brought its market share back to 85 percent in 1986.[25] AT&T has clearly secured a dominant position in this area as a result of its unquestioned technological lead, which allowed it to impose on its competitors the logic of the sheltered culture—a logic that AT&T has fully mastered.

But in the sector of private branch exchanges (PBX), consumer-premise equipment, and rental equipment, AT&T's revenues fell rapidly, from $7 billion in 1984 to under $5 billion in 1986. On the PBX market, only an expensive promotional campaign halted the fall of AT&T's share, which dropped below 20 percent in 1984.[26] Even more serious, in the data-processing equipment and services sector, where AT&T had the ambition to become second worldwide behind IBM, in 1986 the company showed a gigantic loss of $1.2 billion, for total revenues of $2.2 billion.[27]

As the magazine *Business Week* noted, "AT&T has been slow in learning to think like a free-market competitor."[28] Its production costs are too high, and its managers complain of a lack of clear directives. More precisely, said the magazine, what is lacking seems to be a vision capable of mobilizing the firm's 330,000 employees and replacing the public-service objective that had been the central theme of AT&T's management before the breakup. In fact, the great phone company is going through a true cultural identity crisis: a crisis of adaptation to the new universe of the exposed culture.

AT&T and the Cultural Change

The creation in 1984 of AT&T Information Systems, the company's new computer division, marked the end of an era. One of the compensations for AT&T's dismantling was the lifting of a thirty-year prohibition forbidding the company to enter the often-coveted data-processing market. The company was now free to challenge its long-standing rival IBM on its own territory. The computer giant had never hidden its intention to tread on the feet of AT&T, for it also had an eye on the newly integrated computer and telecommunications markets.

At AT&T headquarters there was not the slightest doubt about the success of the enterprise. James Olson, the company's president, declared that AT&T's entry into the data-processing market would "redefine this industry." AT&T already had extensive experience with computer technology. The development of digitally switched telephone exchanges had taught the company how to build high-performance computers. It was no surprise, and it was coherent with the logic of the sheltered culture, that AT&T would attempt to make the best strategic use of this knowledge. AT&T Information Systems decided to market its 3-B series minicomputer, developed by Western Electric for digital switching equipment. This machine also had the advantage of using many components developed by Bell Labs, as well as the Unix operating system (also developed there). Unix is a highly flexible system that can operate on different types of machines. Jack Scanlon, a Bell Labs alumnus who was managing AT&T's computer operations, was counting on Unix to become a universal system and help stimulate computer sales. In practice, however, it was not that simple. Although Unix was widely used in science and engineering, it was unknown in the world of business computing—IBM's territory, and the area targeted by AT&T in its bid to enter the data-processing market. Another handicap for the new product was its price. The 3-Bs were offered at the same price as Digital Equipment's top-of-the-line minicomputers, but they did not have equal performance, and they lacked applications software.[29]

Scanlon's departure after six months as the head of AT&T's computer division led to a profound reorganization of the department and a diversification of the product line. One of the new ideas was to replace some AT&T-built telephone terminals with hybrid terminals able to handle both voice and data. When used with personal computers (the production of which AT&T subcontracted to Convergent Technology), they would enable AT&T's customers to create information and data networks using ordinary phone lines. But this approach failed to take account of two things: first, large number of companies with a need to process and transmit data preferred to use private networks, avoiding the public phone system. Second the characteristics of a computer system, and in particular the availability of suitable applications software, count more in the choice of equipment than the existence of a sophisticated communications and transmission system. Of the 40,000 personal computers AT&T purchased from Convergent Technologies in 1985, no more than 10,000 were ever sold. The others contributed to the $100 million loss sustained by AT&T in the area of hybrid terminals.[30]

One of the successes of AT&T's computer division happened purely by chance. A clause in the agreement signed in 1983 with the Italian firm Olivetti stipulated that AT&T would handle the American distribution of at least 20,000 units a year of an IBM PC clone manufactured by Olivetti. This product was sold as the AT&T PC 6300. Inexpensive and technically well designed, it was an immediate success, and by accounting for 5 percent of the US microcomputer market it prevented the AT&T computer division from showing completely disastrous financial results. Aside from the Olivetti computers, AT&T achieved only modest results in the open market for computers. As table 6 shows, most of its sales in this domain were made to Bell operating companies, which were still captive customers.[31]

AT&T learned at its own expense that it could not manufacture and sell computers as it did large telephone switching equipment. In the old Bell System organization, costs were relatively unim-

Table 6
AT&T computer sales, 1985 ($ million).

Product	Sales to regional telephone companies	Sales to other clients
Microcomputers	30	603
Minicomputers	420	255
Super-minicomputers	650	12
Totals	1,100	870

portant. It was not surprising, therefore, that the prices of most of the products that AT&T tried to sell in unregulated markets were higher than its competitors' prices. When it became involved in the computer business, AT&T possessed not even the most rudimentary analytic accounting tools for following the costs and characteristics of its new products. "There were no statistics for costs, for income, or for customer discounts; no serious product statistics," says a former manager.[32]

The creation of new products was rarely based on a serious analysis of the demand. AT&T's commercial approach was based more on exploiting technical opportunities than on targeting particular customer categories and inquiring into their real needs. In fact, AT&T never gave up its traditional philosophy, which was to produce goods that would then be imposed on a captive clientele. Unfortunately, this sheltered culture approach is not valid in a competitive universe. It leads to unsold products, as AT&T discovered at great cost.

An even more serious problem was that AT&T did not possess suitable personnel or commercial networks through which to reach its potential new customers. As Fortune magazine noted, "Not only did the Unix machines lack applications software to attract company managers and executives, but the salesmen had no means to reach their customers."[33] AT&T's customer base contained essentially names of managers and executives of regional phone companies and of purchasing officers for telephone equipment within companies. Complicating things even further, Judge Green had required that the company separate its long-distance telephone service from its unregulated activities. This led AT&T to establish two separate customer-contact networks. Although it was rescinded later by the judge, this constraint cost the firm a great deal of money.

New Men for a New Strategy

"Will AT&T be able to stay in the computer business?" asked the *New York Times* in early 1987.[34] The company's top management immediately answered Yes, but admitted that the grandiose projects formulated three years earlier were no longer applicable. AT&T will probably never become the equal of IBM, but it still intends to do business at the computer-telecommunications interface. James Olson, who left the presidency to become general manager in August 1986, is interested not only in selling computers but also in selling integrated communications systems. His principal aim is to link the equipment that processes and transmits data in networks, a growing market worth more than $20 billion at the end of 1987 in the United States alone.[35] In this market the major customers are large companies like Boeing, General Electric, and General Motors, which have already begun building large private networks, often with AT&T's support. Such collaboration can often lead to joint ventures in third-party markets. For example, AT&T became associated with Boeing's data processing division to bid on a $4.5 billion voice-and-data-transmission network for the federal government.[36] This is a more familiar universe for the telecommunications giant. Large transmission and data-processing networks are installed to meet the needs of large public and private institutions—both of which follow the logic of the sheltered culture, in which AT&T has always excelled and prospered.

When Vittorio Cassoni took over as head of AT&T Information Systems, in late 1986, this was the logic that he intended to apply. A marketing man and a former IBM employee, he had become the manager of Olivetti's operations in North America after spending six years under Carlo de Benedetti, Olivetti's president. He was therefore used to the logic of the exposed culture. He expressed caution when speaking about AT&T's ambitions in this area. Unlike his predecessors, Cassoni wanted to concentrate on a small number of specific products associating network computer equipment with specialized software. A good example is the system developed by AT&T to enable American Express to improve its handling of slow payers. The system associates a computer with an automatic switchboard. It automatically calls customers who are in arrears, checks to see that someone is on the line, and only then passes the call on to an operator, who will have the customer's file displayed on a terminal. American Express expected this equipment to pay for itself within two years because of the savings in personnel. This example suggests what data processing and new communications techniques can bring to telemarketing, a field in which the market reached $100 billion in 1986.[37]

Cassoni had no intention of limiting AT&T's computer ambitions to a few technological niches at the edge of the sheltered culture. He also hoped to make use of the breakthrough in microcomputing that AT&T had made with Olivetti's equipment. A new-model PC using the latest Intel microprocessor, the 80386 (also used in the IBM PS), was expected to see the light of day shortly. But it would be built by Olivetti, not by AT&T. Everybody should stick with his specialty. The operating system would be derived from Unix but would be produced and marketed with the help of Microsoft, one of the most efficient producers of standardized software, and also a company with experience in the exposed culture.[38]

Cassoni obtained a promise from James Olson that his division would not serve as an outlet for too much "home-grown hardware." He wanted to keep the freedom to acquire saleable products outside AT&T, such as the Olivetti PC. And where is Bell Labs' place in all this? What use is being made of their talent for developing the most sophisticated electronics components? According to Fortune magazine, Cassoni replies to this question cautiously: "Revenue from computers in telemarketing and other niches where AT&T enjoys instant credibility may help fill the chip factories and pay for the Bell Labs. If it doesn't, the sensible strategy would be to scale back the factories and the R&D budget until AT&T has learned how to market."[39]

The logic is clear. Cassoni is a professional of the exposed culture who has erupted into another universe, which he wants to change. "Adapt or perish" is the law in the world he comes from, a world with which AT&T is now confronted. Many feel that such adaptation is impossible. In the best of cases the cost of such adaptation may be very high, not only for the company but also for the entire United States.

Who Benefits from Deregulation?

Washington is a center for confrontation between interest groups in American society; it is the place where their values or dogmas find expression. The dogma of deregulation, which appeared in the 1970s, was only one component of a powerful wave that was to flow over America for ten years: the "less government" wave. The pendulum movement that during the late 1960s had brought the interventionists to power, supported by the environmentalists had reached its apogee. The economic crisis provoked by the first oil shock had served as a revelation, turning the spotlight on the need for greater economic efficiency in a world that was increasingly open to international competition. In this context, the overwelming influence of government was increasingly being criticized as one of the reasons for the loss of America's competitiveness. "Less government" had already become a political slogan well before the election of Ronald Reagan. The deregulation of a number of public services, such as telecommunications, ground and air transportation, and electrical power, seemed to be a good starting point for turning this slogan into reality. Under regulatory shelter, a number of monopolies, oligopolies, and cartels had flourished. In contrast, deregulation would allow new firms to enter these formerly protected markets. Prices that had been fixed and controlled by the government would now move freely according to competition and market mechanisms. Consumers would be the first beneficiaries of this new competition. After being introduced in telecommunications in the early 1970s, deregulation was applied to railroads (1976), airlines (1978), and later to trucking and banking services.[40]

A decade later, the results of this unprecedented upheaval in important sectors of the American economy showed great contrasts.

From the consumer's point of view, deregulation had produced some unquestionable benefits. The most visible of them was the drop in prices on high-traffic routes, whether in air transportation, telecommunications, or railroads. Intense competition arose for

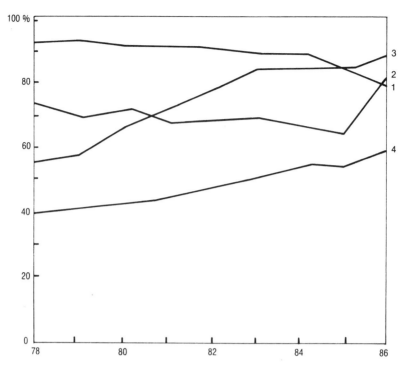

Figure 9 Concentration in deregulated services. (1) Long-distance telephone service (AT&T's market share); (2) air transport (market share of six top companies); (3) rail transport (six top companies); (4) road transport (ten top companies). Source: *Business Week*, December 22, 1986, and other sources.

market shares on the most profitable routes. But this result was often produced at the cost of poor local service or increased prices for it (as in telecommunications) or the disappearance of the less profitable routes (in rail and air transportations). Furthermore, the lower profit margins stemming from competition often had an impact on the quality of service. For example, in air transportation, less attention to maintenance is one of the reasons for the increasing number of delayed or canceled flights. Many specialists are also concerned about the potential implications for passenger safety.[41]

After an initial phase during which the entry of a large number of new firms could be witnessed in deregulated sectors, the industrial structure again evolved toward concentration (figure 9). Free competition confers a clear advantage on well-established companies, which can benefit from economies of scale derived from the extent of their networks. For instance, telephone companies competing with AT&T found it difficult to compensate for the handicap of their smaller networks.[42] In the air transportation field, a large number of small companies were absorbed as the result of bankruptcy or financial problems, and there was a reconstitution of regional quasi-monopolies. Northwest Airlines, which in 1979 had carried 40 percent of the traffic in and out of Minneapolis-St. Paul, carried 80 percent in 1987. It also controls 87 percent of the traffic at Memphis, and 65 percent at Detroit, after acquiring local companies.[43] In order to lower their costs, the big airlines tend to organize their networks around one or two hubs, where they rapidly achieve a dominant position. Not only do passengers no longer have a choice in getting to these cities, but they become mandatory transit points for most of the company's flights, which extends the practice of indirect flights. This new organization also partly explains the deterioration of service. The late arrival of a single airplane may produce a chain reaction when travelers are required to pass through the company's hub and to make connections there.

In the field of equipment supplies, adapting to the competition has most often been accomplished at the expense of national producers, which had previously enjoyed sheltered markets. By opening equipment supply to competition, deregulation chiefly benefited foreign manufacturers. This result is unquestionable in telecommunications, where the consequences of deregulation were amplified by the dismantling of AT&T.

"Dismantling of the Bell System opened the American telecommunications market to a flood of imports," noted the *New York Times*.[44] This result, even if it surprised most observers by its extent, was in part predictable. It illustrates what may happen when an administration applies sound ideas too blindly, for the sake of principles, without always anticipating the consequences of the decisions. This dogmatism was compounted by ingenuousness when the same administration, seeing the penetration of foreign products into the US market, asked other countries to adopt the same approach to opening their borders, under the threat of reprisals or the erection of similar protectionist barriers.

Deregulation, increased competition from abroad, and technological change all work, therefore, to narrow the extent of the sheltered culture. After the euphoric period during which previously sheltered firms such as AT&T and Boeing were the first to acclaim the new rules of the game, it turned out that such companies encountered serious difficulties in adapting to these new rules. Meanwhile, competition became increasingly fierce within the exposed culture, to the point that the retreat of American firms in certain industries (e.g., computer memories) raised serious questions about the long-term future of those industries.

Is this evolution irremediable? Is this retreat on all fronts of American high-tech firms irreversible? Or are there signs to suggest that adaptation may be taking place? This is the question to which we will now turn.

Made in USA

11

The National Bureau of Standards (NBS), located in Gaithersburg, Maryland, about 15 miles from Washington, has a special status among the large federal laboratories. Aside from its research activities, it has the job of creating and testing technical standards for industry. This was already important in the area of mechanical and chemical development, but the NBS's task is even more significant today as information technologies spread throughout the industrial world. One of its more advanced programs in this area involves a pilot facility for research in automated production, the Automated Manufacturing Research Facility.[1]

Automating factories is nothing new. For 20 years numerically controlled machines and robots have been gradually coming into use in various industrial sectors. Their use explains a large part of the gain in productivity recorded during this period. More recently, the widespread use of computers have paved the way to a higher level of automation. This is the purpose of the NBS's pilot facility. It includes several work stations that are fully automated for handling various manufacturing operations, programmed and controlled via a minicomputer. This makes it possible to program—from a site outside the shop, and without any worker assistance—all the operations required to manufacture certain pieces of equipment. The programming covers tool changes as well as the movements of parts and materials to and from storage areas.

Associated with a computer-assisted design (CAD) unit and an automated system for handling and managing inventory, this computer-assisted manufacturing (CAM) equipement has been developed into a completely integrated automatic manufacturing system, known as computer-integrated manufacturing (CIM), that is changing the face of American industry.[2]

The most important aspect of CIM is the control and distribution of the flow of information in a company. CIM connects and integrates in a single computer network all of its equipment and functions, from product design to manufacturing, including purchasing and inventory management. This is more complex than it might seem. For example, in most companies already equipped with a CAD system and with numerically controlled machine tools, the data coming from the CAD system cannot be directly used by the machines in the production shop, which have their own languages and communications protocols. It is a real headache to connect machine tools and computers of different makes and designs; it requires powerful software and sophisticated communications protocols. By accomplishing this integration, CIM provides global optimization of the production units, including input and output of materials, spare parts, and finished products ready for delivery.

Minimizing stock is just as important as increasing efficiency in the manufacturing process. By lowering overall labor costs, CIM makes it possible to manufacture products at lower cost. Surprisingly, however, the most significant gains do not relate to manufacturing as such (for which, in the United States, labor accounts for only 10–15 percent of the total cost of products), but to indirect labor costs (which account for 45 percent). Included here in particular are the costs of supervisory and management personnel, maintenance, and the movement of products and information from one department to another. Significant savings are also obtained with respect to another function, one that represents 25–30 percent of the cost of traditional manufacturing: quality control. CIM leads to products that are not only less expensive but also of better quality. Its chief advantage, however, is its flexibility. Computer-controlled manufacturing makes it possible to program numerous variables for adjusting to changing demand, and for switching from one product to another. In fact, CIM is creating a revolution in manufacturing methods. It makes it possible to produce small series of products at prices that previously could be achieved only with large installations. It compensates simultaneously for the advantages of economies of scale and for those of producing goods in countries with low labor costs. CIM is the key to bringing factories back in the industrialized countries. It is one of the possible responses to the superiority of products imported from the Far East. For American firms, it is one means of reconquering the exposed culture.

IBM Brings Its Factories Home

Aside from the need to improve productivity, there is another important reason for IBM's extensive use of these new techniques of automation: securing the supply of strategic components. A few years ago, because of lower cost, the company bought most of the memories needed for its 3081 mainframe outside the United States. Today all the memories used in one of the most recent models of this series, the 3090 Sierra, are manufactured in IBM's own components factory in Burlington, Vermont.

In 1985 IBM invested $3 billion to modernize its production tools. Between 1986 and 1989, it planned to spend more than a third of its total investment budget for this purpose—nearly $50 billion over that period.[3] In the early 1980s, when it had just begun marketing its personal computer, IBM was buying most of its components abroad, mostly in Southeast Asia. Today they are manufactured in the United States, along with most of the peripherical equipment. Disk drives are produced in an ultramodern plant in Rochester, Minnesota, and keyboards are made in Lexington, Kentucky. Printers, which IBM had been buying from the Japanese manufacturer Epson, are now being built in Charlotte, North Carolina. IBM insists that it has the most efficient printer factory in the world. According to Raj Reddy, director of the Robotics Institute at Carnegie-Mellon University, in 1986 this plant was using probably the most advanced computer-driven production system available.[4] Put into service in 1985, the plant uses 50 robots, 160 computers, and 200 workers. New product design has made it possible to simplify the product, reducing the number of parts to be made from 150 to 60. This simplification went hand in hand with an increase in performance from 80 to 200 characters per second.[5]

IBM is not the only large company in the American exposed culture to follow this course. In 1985 Apple opened a fully automated production unit in Fremont, California, to produce McIntosh machines. The plant uses 250 computers and employs only 180 persons to produce 1,200 computers a day.[6]

During the decade following the first petroleum shock, American multinational corporations gradually relocated an important part of their production to countries with cheap labor, taking advantage of the worldwide opening of markets. It seems, however, that a reverse trend has begun to bring production units closer to their markets, in particular on American territory.[7] The new manufacturing technologies, which obviate the earlier comparative advantages of cheap-manpower countries, are among the reasons for this movement.

The General Motors Protocols

Sales of automation equipment, including robots, computers, and communications equipment, doubled between 1980 and 1985, reaching $18 billion. This figure was expected to double again by 1990.[8] The big companies are the ones most active in this area. According to the Yankee Group consulting firm, 90 percent of CIM investments will be made by fewer than 2 percent of American companies over the next years. In addition to IBM, some two dozen firms in aeronautics, computers, electrical equipment, and automobiles have already become pioneers in this area. The most strongly committed are the automobile manufacturers. Automation of factories has been, for them, a matter of survival. General Motors alone has invested $40 billion in highly automated factories since the beginning of the 1980s.[9] The Detroit firm has paid particular attention to defining its communications standards. Its objective is to be able by 1990 to link via networks most of its computers and programmable systems—some 40,000 in all—and those of its suppliers as well. The result has been the MAP (Manufacturing Automation Protocol), specially designed for handling and transmitting the data involved in a manufacturing system with highly integrated automation. Designed to handle 10 megabits per second, this wideband protocol is in accordance with the seven-level OSI (Open

System Interconnection) standards. It defines the rules for acquisition, transmission, and processing of information, as well as the control modes between different pieces of equipment connected to the network (particularly computers and machine tools). MAP is becoming widely used in automation. Several large manufacturers of CIM equipment, including Honeywell and Hewlett-Packard, have adopted it, as have large users such as Eastman Kodak. In 1986 more than 400 vendors of automation equipment committed themselves to providing product compatibility with MAP.[10] General Motors assembled the expertise needed to succeed in entering this era of information technologies by acquiring the military electronics firm Hughes and the data processing consulting company Electronic Data Systems in 1984.[11]

Although the high cost of CIM now limits its use to a few large companies, its largest potential for application lies with small and medium-size companies. Cypress, a manufacturer of integrated circuits in San Jose, California, is a good example. It illustrates the benefits a small company can derive from the use of modern manufacturing methods. Cypress, created in 1983, invested $25 million, most of which went toward installing a fully automated production line with twenty robots. The line, which takes advantage of the flexibility of computerized manufacturing systems, produces seventy different models of integrated circuits. Cypress' advantage over its Far Eastern competitors lies in a production mode that guarantees better product quality at competitive prices and allows for the production of circuits that meet the customers' requirements exactly. The company's income tripled in two years, and in 1986 it reached $55 million.[12]

For each small company like Cypress, however, there are a hundred others that are still ignorant of CIM's possibilities. These companies are the main target of the NBS pilot plant. One of the objectives of the program is to familiarize small companies with new methods of automation. The Department of Commerce, which controls the NBS, estimates that 75 percent of manufactured products sold in the United States come from approximately 900,000 small and medium-size companies, including a large number of subcontractors in the aeronautics, automobile, and military sectors. Until now, most of them have not taken part in the great movement toward automation. For their benefit, the Department of Commerce has attempted to popularize a new concept, the Flexible Integrated Manufacturing Center. This is a production unit that is totally controlled by computer, and that can operate around the clock for a group of small companies. The high level of automation allows it to be rapidly adapted to the production specifications of the various participating firms, at an attractively low cost. The concept seems destined for a bright future, for it provides an answer to the financial barrier represented by the initial cost af CIM. The first Flexible Integrated Manufacturing Centers have gone into services in Grand Rapids, Michigan; Meadville, Pennsylvania; and Troy, New York.[13]

To assess these developments, it is useful to weigh them against what is happening abroad. Japan is also making a major CIM development effort.[14] But I should emphasize the extent of the effort being made in the United States in this field—an effort that, if continued, will produce the greatest transformation of the American production system in 40 years. This adaptation to a new manufacturing logic is of particular importance in the manufacturing of computer memories.

Computer Memories: Halting the Japanese Invasion

Sematech would finally set up shop in Austin, Texas: this was the decision announced on January 6, 1988, by Charles Spork, president of National Semiconductor and president of the consortium of producers and users of integrated circuits that joined forces to create Sematech.

To attract this trailblazing project of the American integrated-circuits industry, the state of Texas had gone so far as to purchase facilities from Data General—including a "clean room" adapted for producing silicon wafers using submicronic technology. In all, Texas invested $35 million, with an equivalent sum made available in the form of low-interest loans, fiscal exemptions, and various local benefits. However, Texas' major asset was probably Jim Wright, the Speaker of the House, who did not hide his intention to use his influence to make sure that Sematech would benefit from all the federal financing the project would require—on the condition that the consortium would set up shop in Austin! Lloyd Bentsen, chairman of the Senate Finance Committee, also played a part in this. Locating Sematech in Texas would reinforce the state's electronics industry, which was already important. Aside from having a large number of military electronics firms, Texas has become the third-ranking region in the United States in the production of integrated circuits, after California and the Northeast. Texas Instruments, the second-largest American producer, is headquartered in Dallas and has major production facilities there. IBM, Motorola, Advanced Micro Devices, and the Franco-Italian group SGS-Thomson have plants in Texas as well. And Austin is the headquarters of the Microelectronics and Computer Corporation, a research consortium created in 1983 by twenty American companies.[15]

Up to the day of the decision there was a battle among the half-dozen states best qualified to attract Sematech. At the last minute, Massachusetts improved its offer, bringing it to nearly $400 million in the form of loans, interest rebates, and fiscal exemptions; its proposal also included the possibility for Sematech to be housed at the brand-new Massachusetts Microelectronics Center. Arizona also held a trump card: ultramodern facilities that previously belonged to SGS, representing a $19 million investment, would have been placed at the disposal of the project for a nominal fee. Loans, fiscal exemptions, and access to the laboratories of the University of Arizona completed the proposal, estimated at $184 million.[16]

Sematech represents an important gamble not only for the American microelectronics industry but also for the region hosting it, since it could expect to stimulate other industries.

Sematech's objective is to develop advanced microelectronics manufacturing technologies, in particular those required for producing new generations of memories—16- and 64-megabit DRAMs and 4-megabit SRAMs. This is a field in which, as we have seen, American firms have progressively given up fighting their Japanese competitors. The program's first phase will deal with microlithography techniques needed for manufacturing circuits with features 0.8 micron thick. The second phase is to make it possible to go down to 0.5 micron. The third phase is expected to culminate in

1994 with the industrial application of 0.35-micron technology. Thinner features are the means to increasing chip density.[17]

The heart of the project includes several fully automated pilot production lines, on which tests will be made not only of these microlithography techniques but of all the generic technologies needed for producing different types of circuits. The essential objective is not to carry out avant-garde research, but to develop integrated manufacturing systems using the most recent technologies and to test them in an environment similar to the factory environment in which they will be applied.

Such a project is costly: $1.5 billion over six years for the first three phases. This is the main reason why the consortium was created: to pool the resources of the corporations concerned. The participants include the major American manufacturers of integrated circuits—Motorola, National Semiconductor, Intel, Advanced Micro Devices, Texas Instruments—as well as IBM, Digital Equipment, and Hewlett-Packard, which produce circuits for their internal needs. The consortium expected government assistance for 50 percent of the total expense. At the end of 1987 Congress approved a $100 million subsidy, taken from the 1988 Department of Defense budget, for the first phase of the project.[18]

Even though disagreement persisted in Congress in 1988 over the specific conditions for this contribution, its acceptance in principle was an event in itself. This was the first time that industries of the exposed culture would have the benefit of significant direct government aid. The fact that this aid has been officially justified by the Pentagon's fear of seeing strategic domestic manufacturing capabilities emigrate abroad should not be misleading; for the first time the American government is directly providing funds to improve the productivity of firms in the exposed culture that are threatened by foreign competition.

This direction was the result of a very thorough analysis. Different groups of experts who, during 1986, examined the competitiveness of the US integrated circuits industry all came to the same conclusion: only an unprecedented effort in manufacturing techniques could guarantee a future for this industrial sector.[19] Unless this were done, the production of all standard circuits would eventually be left to the Japanese. In its own way, Sematech represented a burst of energy from an industry that was as strongly threatened as the automobile industry had been ten years earlier. It is unlikely that the mastery of new manufacturing techniques can alone solve all the competitiveness problems in the sector; but such mastery does promise to provide, as it did in the case of automobiles, a much-needed breath of fresh air for an industry threatened with suffocation.

What is at stake with Sematech is a demonstration of the United States' ability to adapt and respond to the foreign invasion in a particularly threatened sector of the exposed culture; hence the highly symbolic value of this project.

The Virtues of "Human Investment"

The CIM revolution is not only technological. It also involves the organization and the management of companies. It foretells an end to large factories, to the benefit of small production units. It represents the emergence of a new school of management to replace Taylorian factory organization.

The type of organization proposed at the end of the last century by Frederick Taylor was based on breaking down and rationalizing tasks. Elementary tasks were dissected, timed, and optimized one by one, leading to greater production efficiency. This also made it possible to use less skilled labor, since segmented tasks required less know-how. Technical progress was an essential ally in this approach. Every task that could be performed by a machine rather than by a worker meant a further step toward a more efficient form of industrial organization. These principles were triumphant in the creation of the first assembly lines at the beginning of the century.

The limit to automation is the complexitiy of certain tasks that only workers can and should continue to perform. But as technology evolves, there are fewer of them. As a result, switching and adjusting tools—a task long performed by workers when manufacturing mechanical parts or when changing from one series of parts to another—is now being done automatically, using computer programs. This involves numerically controlled machine tools, the use of which began during the 1960s. The worker's or technician's know-how was replaced by a computer program. Thus, at first, the appearance of computers in production only reinforced Taylorian logic. Computers displaced intelligence from the shops to a central programming and controlling department. Over the years, however, this logic evolved. Its alienating character was criticized by the unions, and in any case it proved unable to resolve what had become a major problem for industry: declining product quality in large production lines. Its global efficiency therefore came to be questioned, as it was seen that investments in automation were becoming less profitable. The quest for greater efficiency, it was felt, should henceforth increase the workers' motivation and reduce the repetitiveness of tasks. Experts in factory management and organization should henceforth focus their attention on the worker and on his working environment and constraints. Nevertheless, the underlying philosophy remained the same; in spite of appearances, the man-machine relationship had not changed. Man had a role to play only because there was nothing more efficient in sight in a system that tended progressively to exclude him.

True change came about with what were known as the "new managers." *In Search of Excellence*,[20] Thomas Peters and Robert Waterman's analysis of the most brilliant American companies, became the bible of management. What do we learn here? That the most successful companies are the ones that know how to motivate their employees and define a collective objective with which each can identify; the ones that are able to create a team spirit that leaves room for individual expression at every level of the company's organization. Peters and Waterman give the example of Dana, a Midwestern company selling automobile gearboxes and spare parts. In the late 1970s, Dana called attention to itself by showing up in second place in *Fortune* magazine's classification of the most profitable companies. In the early 1970s, sales per employee had been about the same for Dana as for the rest of the industry. But within a few years they had tripled, without any sizable investments by the company. For Peters and Waterman this phenomenal gain in productivity was the result of an unusual effort to increase employee motivation and improve the quality of human relations within the company. Yet the company was highly unionized, dominated by the powerful United Auto Workers. Dana had nonetheless become a model of "participative management," a strategy that associates management, unions, and stockholders in what amounts to a new sharing of power in American companies. Lester Thurow sees in the widespread application of this type of organization a condition for a renaissance in American industry. In order to develop the incentives and teamwork needed for a prosperous society, he says, America should change its basic conception of a firm. It should no longer be a battlefield between management and unions but an association between workers, management, and stockholders, with the common goal of maximizing the company's production.[21]

Another factor in the best firms, shown by Peters and Waterman's analysis, is the size of production units: "small is beautiful." Small units, with 500 persons at most, most often lead to higher productivity. The most efficient companies are the ones that decentralize management, giving broad responsibilities to operating units with maximum turnover of $30 million to $50 million. This mode of organization could explain the success of firms as different as Texas Instruments, Hewlett-Packard, 3M, and Johnson & Johnson. Small size and decentralized responsibility are in fact the essential conditions for motivating men, say Peters and Waterman. They are also the conditions that make it possible to use the new automation potential of CIM.

Philippe Messine, in his book on automation in the United States, says: "The entrepreneurs that are the most aware of the limits of Taylorism are searching for another type of relationship between man and machine. They understand that robots cannot replace men, that flexible production shops cannot replace teamwork, and that artificial intelligence cannot replace the human brain. Instead of technology as a 'substitute for man' they prefer to see it as an 'extension of man,' able to increase his productive capacity."[22] Human investment and job requalification go hand in hand with the higher forms of automation in such a decentralized organization of production. The large amount of money spent on training is one of the common characteristics that Peters and Waterman found among the best American companies.

The consensual management model, imported by Japanese firms starting to do business in the United States, also has had an influence on this new American school of management.

"Japan, Inc." Arrives and Takes Root in America

Things have changed greatly since the day in 1982 when the "CBS Evening News" showed a group of Detroit auto workers smashing Toyota windshields with sledgehammers while denouncing with anti-Japanese and anti-Asiatic slogans the people they held responsible for the massive layoffs in the American automobile industry. By 1988, Japanese firms had become creators of jobs on American soil. The first ones were automobile manufacturers. In 1982 Honda set up shop in Ohio, and was soon followed by Nissan in Tennessee and Mazda in Michigan. A little later Toyota, the largest Japanese manufacturer, installed a facility in Kentucky, thus adopting the same strategy of substituting some of its exports to the United States with products manufactured locally.[23] The strategy was all the more profitable in that the yen's value relative to the dollar had been rising continuously since 1985, making direct exports to the United States expensive. Toyota's objective was to produce 1.6 million vehicles a year in the United States by the 1990s.[24]

A similar phenomenon occurred, beginning in 1985, in the electronics sectors. NEC invested in an optical fiber plant in Portland, Oregon. In May 1986, Matsushita announced its decision to build two US factories, one in Washington State to make television sets and the other in Georgia to build high-fidelity components. Toshiba decided to build a video recorder plant in Tennessee, and Sony expanded its color TV facility in San Diego.[25]

By the end of 1986 there were 600 Japanese factories in the United States, employing 160,000 persons.[26] Japanese sources estimate the number of jobs that Japanese firms could create in the United States by the year 2000 to be 850,000.[27] Buying bankrupt or declining American factories in the "Rust Belt," the Japanese firms have been warmly welcomed by state authorities, who see their projects as a way of revitalizing local economies. The Japanese have imported their concept of work organization while calling largely upon local resources. The examples in the auto industry show that within a few months the Japanese managers completely transformed work relationships in factories previously managed by Americans, with spectacular increases in productivity. When asking for financial sacrifices by the workers, they began by reducing their own salaries. They set up new channels of communication within the company and encouraged individual initiative. In a number of cases, they have had the active cooperation of Americans unions.

One of the best examples of the value of the Japanese methods is a factory in Fremont, California. This plant was involved in the NUMMI (New United Motor Manufacturing, Inc.) project, in which Toyota and General Motors jointly manufactured a compact car, the Nova. The Japanese put $150 million in the pot, while the Americans put up $20 million, along with the facility and 5,000 workers. But what a facility! "It was a highly developed example of everything that could be wrong with American management. Absenteeism was raging at about 20 percent. The union had about 800 grievances pending. Quality and productivity were abysmal. Drug and alcohol abuse were rampant. Fights would break out on the floor. Labor-management relations there, says local UAM head Tony DeJesus, meant constant preparation and training as to how to fight with each other."[28]

Now this factory, a model of participative management, can hardly be recognized. The atmosphere is relaxed. Each individual seems actively involved in *kaisen,* a Japanese word that all the workers learned to pronounce and which means a constant striving to improve each person's efficiency. The old factory's burdensome and complex hierarchy has disappeared, along with the union's chronic antagonism toward it. One of the major innovations was *adon,* which means that any worker has the right to stop the production line if he deems it necessary. It was difficult for American supervisors to accept this innovation, which had symbolic value in the eyes of the Japanese managers.[29]

Will the Japanese–American honeymoon last? It is still too soon to say. In any case, an industry of a new kind, dynamic and efficient, is making its appearance in certain areas of the United States.

A New Regional Dynamic

The image of a "Rust Belt" of declining or disappearing traditional industries no longer corresponds to reality. Today the traveler to Pittsburgh, the former capital of coal and steel, will find a clean and prosperous city whose local specialties include software, robots, and artificial intelligence. This is the result of an active conversion policy implemented by the state of Pennsylvania and the city of Pittsburgh, and the consequence of the spread of high-technology industries to new regions of America.

The decline of the steel industry struck the Pittsburgh area hard. A local organization, the Allegheny Conference on Community Development, was established to deal with the situation. It played an essential role in rehabilitating the urban fabric, with financial assistance from the state of Pennsylvania, the city, and the region's chief economic agents, particularly the banks and large companies such as US Steel, Alcoa, and Westinghouse. At the state level, assistance was also forthcoming for industrial conversion and training programs. This project was amplified during the 1981–82 recession through the creation of the Economic Revitalization Fund, with a $190 million endowment. A broad public debate on the "choices for Pennsylvania" made it possible to launch what we in Europe would portray as a true state industrial policy.[30] The aim was to stimulate the growth of new industries in order to compensate for the regression of the declining sectors, and to lessen the impact of this decline with appropriate subsidies and policies, particularly for manpower conversion and training. Like most other states, Pennsylvania actively sought outside investments. Large industrial projects such as automobile plants have always been the subject of competitive bidding among states. State incentives worth tens or hundreds of millions of dollars are common in such cases, as was noted in connection with the Sematech project. When they are back home and close to their constituents, congressmen generally adopt a much more pragmatic attitude with respect to public subsidies for industry than they may take when deciding whether to allow federal intervention in this domain. Pennsylvania is only one example; Ohio, Michigan, Massachusetts, and many other states have developed similar industrial policies.[31]

Within this framework, many states have given top priority to high-tech industries. These are generally the fast-growing sectors. While they may create only a small number of jobs directly, indirectly they generate many—particularly in services.[32] More than anything else, they may contribute to the creation of self-sustained local industrial development. The models of this type of growth are, of course, Silicon Valley and Route 128, which many states have tried to duplicate. In their endeavors, local governments have developed policies that attempt to compensate for local scientific or industrial inadequacies or weaknesses.

Among the first of these regional and local actions have been the reinforcement of the university system and the development of close relationships between universities and industry. Twenty years ago, university-industry ties were limited to a few establishments, such as Stanford and MIT. But since the mid-1980s much state financing has been made available to strengthen such relationships. A growing number of universities, concerned with finding jobs for their graduates and diversifying their financing sources, have sought to develop research programs in collaboration with industry. This convergence of interest has given birth to a broad array of new regional and local organizations to stimulate innovation, industrial research, the creation of new firms, and technology transfers from universities to industry. In many states these local and regional initiatives have favored the birth or the development of a new growth dynamic centered on high-tech industries.[33]

In mid-1987 there were 140 "incubators"—centers hosting new firms—where entrepreneurial candidates could find inexpensive offices and technical, legal, and financial assistance for launching a project. There were also some 50 centers facilitating the transfer of university research results to small regional companies. And there were some 100 research parks, usually near campuses, used by the more developed companies to maintain close ties with university laboratories.[34] Certain states also created venture-capital funds sustained by public financing. These are intended to fill the gap left where private funds are inadequate for creating or developing high-technology companies, as happens in certain states. Venture capitalists in fact tend to concentrate their activities in the traditional high-technology regions—California and the Northeast.[35]

The states are also generally concerned with reinforcing the regional applied-research infrastructure. They often do it by sponsoring new research centers in which rival companies in a given sector are associated with university laboratories in the same specialty. Stanford University's Center for Integrated Systems, as we have seen, illustrates this new type of cooperation. Such centers have multiplied in recent years. In mid-1987 there were more than

120 of them.[36] In spite of their great diversity, they have a number of essential characteristics in common.

First of all, the states and the universities contribute to their financing, along with federal agencies and private firms. Some centers were created by groups of companies in order to pool resources and to share the cost of pre-competitive research.

Their second common characteristic is their specialization. Composite materials, computer graphics, genetic engineering, microlithography, and automation are examples of the specialties they cover. Each center is devoted to the study and development of a particular range of "generic technologies"—basic tools potentially useful to all companies. These technologies are applicable not only in the high-tech industries but also in a large number of traditional industries. The function of these centers is to "irrigate" not only the high-technology companies in the exposed sector, but the entire industrial matrix that is potentially able to use these technologies.[37]

Their third common characteristic is that these centers offer students an opportunity to work on industrial problems, thus preparing them to enter the working world. The companies derive double benefit from this: it provides them with an inexpensive and competent work force; and it enables them to identify the more promising students, whom they may hire later. Access to a supply of specialists has become a major concern for high-technology firms. A shortage of qualified personnel will probably create for certain companies a bottleneck that will limit the spread of new technologies throughout American industry over the next ten years. This explains the attention being given to training young people through research in cooperation with industry. The $30 million spent by the National Science Foundation annually since 1984 to develop its cooperative Engineering Research Centers (ERCs) reflects this priority.[38]

Of course, the growth based on high-technology industries that has been apparent since the early 1980s in a number of states often stems mainly from spontaneous evolution. High-technology industries—particularly those of the exposed culture, such as electronics and computers—have spread around the country, away from the geographic areas where they were born.[39] This scattering has occurred both in regions with an old industrial tradition, such as Ohio and Pennsylvania, and in southern and western states, such as Florida, Texas, and California. Silicon Valley and Route 128 are no longer the exclusive lands of high-tech industry; as it has spread across state borders, the original model has adapted to the regions hosting it. At the beginning of 1987 the magazine *High Technology*[40] counted thirty geographic areas—a dozen of which were in traditional industrial states in the East and the Great Lakes region—that were giving birth to high-technology metropolises, with their own growth dynamics (see figure 10). Each one of these new high-technology meccas had one or more technological specialities (e.g., robotics in Pittsburgh, biomedicine in Baltimore); and each possessed one or more universities, research parks, or institutions created specifically to stimulate technology transfer.

The excellent symbiosis between universities and industries, and the swarm of companies it has led to, has always been one of the great assets of the US development model for high-tech industry. The spread of this model to a large number of states outside California and Massachusetts, with the support of vigorous regional development policies for technology and industry, is a real asset for propelling a large portion of the country toward the logic of the exposed culture.

The Brain Drain

The American scientific system is still the world's most outstanding, and continues to attract the best specialists from around the globe. A planet-wide brain drain pulls the top talent to American universities and high-technology firms. In the universities, 40 percent of the doctoral candidates in the scientific and engineering disciplines are foreigners. This figure rises to 52 percent in electrical engineering and computer science. In 1985, 55 percent of the doctorates awarded in these disciplines went to foreign students.[41] Though many of them later return to their home countries, a significant proportion establish themselves in the United States. This is particularly true for many Asiatic, Indian, and Iranian researchers and engineers, who are unable to turn down the attractive offers made to them by American laboratories. Table 7 shows

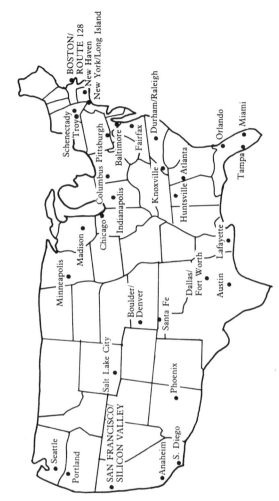

Figure 10   The new high-technology metropolises.

Table 7
Immigration to the United States by foreign engineers and scientists, 1985.

Country of origin	Number
India	1,552
Taiwan	1,044
Great Britain	979
Philippines	713
Canada	687
Iran	672
West Germany	301
France	131
Japan	87
Other	4,814
Total	10,980

Source: NSF 87–319

the distribution of these new immigrants by country of origin. The numbers represent applications for permanent resident status—a necessary step toward American citizenship—accepted by the US government in 1985 from highly skilled engineers and scientists.

The numbers in table 7 represent only the researchers and engineers who have decided to stay permanently in the United States and have been authorized to do so. They do not include all those—and there are many—who have long-term visas that they manage to renew on a yearly basis without requesting permanent resident status, or those whose applications have not yet been accepted. A second observation has to do with the European countries. Western Europe annually loses about 2,000 engineers and scientists who emigrate permanently to the United States. This is a troubling hemorrhage. European researchers do not have the excuse of many Third World scientists, who cannot find a scientific infrastructure or an environment enabling them to carry on their work in their own countries.

The figure of 11,000 immigrants a year may seem small in comparison with the total population of American engineers and scientists—3.2 million in 1984, including 1.2 million scientists and 2 million engineers. But the quantative aspect is not all that counts. These new American citizens are unquestionably members of the scientific and technical elite in the countries they come from. Often, as we have seen, it is the best who are attracted by the exceptional career opportunities they can find in the United States. This is reflected in the number of scientific awards that have been won by naturalized American researchers. Since 1950, they have won 14 out of 45 Nobel Prizes in medicine, and 12 out of 31 in physics. Some recent examples are the 1983 Nobel Prize in physics, awarded to Subrahmanya Chandrasekhar, originally from India; the Nobel Prize in medicine awarded in 1981 to David Hubel, from Canada, and Torsten Wiesel, from Sweden; and the Nobel Prizes in chemistry awarded in 1981 to Roald Hoffmann, from Poland, and in 1983 to Henry Taube, from Canada. To this list should be added the two Nobel Prizes in economics awarded to Americans of European origin: the Frenchman Gerard Debreu (1983) and the Italian Franco Modigliani (1985).[42] In addition to their scientific skills, these immigrants, who are contributing to sustaining the brain drain, are bringing to America another important asset: their strong motivation. As with their immigrant predecessors of all conditions and origins, one of their main ambitions is to become Americans—to integrate themselves into the country that has welcomed them. In order to do so, they are prepared to give the best of themselves. America is still seen as a land of opportunity by many who wish to take risks and succeed.

A Double Adjustment in the Sheltered Culture

Changes are also taking place within the sheltered culture—changes that affect the behavior of firms as well as government programs.

The main change taking place in companies is a result of international competition. Firms such as AT&T and Boeing must contend from now on with Northern Telecom and Airbus Industrie. Competition introduces new rules of the game in the sheltered culture. At the end of the 1970s, Boeing was obliged to expand its marketing department and notably improve its productivity. An expert in the aeronautics industry noted in late 1987 that the Seattle firm had adopted an "Airbus style" of marketing and was paying more attention to commercial promotion, including new incentives for customers to buy or lease its aircraft.[43] In fact, in some segments of the aeronautics industry today, competition is as intense as in many industries of the exposed culture. A similar revolution is underway, as we have seen, in certain segments of the telecommunications industry.

Even if they are expensive for the firms confronting them, these changes are, from one point of view, salutary. They are introducing greater efficiency in a universe where such matters were once often neglected or even ignored. They do not necessarily imply a change of culture. Even though competition has changed some of the ground rules for selling telephone switching equipment to the Bell operating companies, AT&T is still faithful to the behavioral logic that it always employed, using its technological advantage to resist its competitors. "Competition by technology" remains the most common practice in the sheltered culture. The same can be said of Boeing in the market segments that this American manufacturer still dominates completely, such as that for long-distance jumbo jets. Other indices seem to show that Boeing still has its roots in the sheltered culture. Its decision to take an equity position in one of its large customers (United Airlines) reflects the logic of the sheltered culture,[44] as does the development of its defense division during the first half of the 1980s. Boeing's military activities increased from 17 percent of the company's total revenues in 1980 to 40 percent in 1984.[45]

Increased competition, therefore, does not necessarily cause a company to swing into the exposed culture. This should not be surprising, since the distinction I have made between the two cultures, and between the two corresponding innovation processes, is linked to specific characteristics of the products themselves and to the structure of the demand. Boeing's development and sales of a new jumbo jet remain tied to the needs and specifications expressed by a few large airlines, just as for AT&T the characteristics of new digital telephone equipment depend on the particular requirements expressed by a few regional telephone companies. Large telephone switching equipment is still a complex product that must be adapted to the specific needs of each customer. It cannot be designed or manufactured like consumer-premise telephone equipment, which is mass-produced in millions of identical copies. Without changing its nature, the sheltered culture is being transformed under the effects of increased competition as the firms adapt to a new situation.

Some consequences of these changes are perceptible in Boeing's global strategy. Its decision to temporarily abandon development on the 7J7, after spending $100 million on it, reflects a reevaluation of its objectives and priorities. By concentrating on modernizing its 737s and 747s instead of bringing out a new model, the company is making resources available in the short term to reinforce its commercial activity.

In September 1987 the *Wall Street Journal* noted that cancellation of the decision to develop the 7J7 would enable Boeing to devote some $500 million in 1988 to its increasingly difficult marketing battle against Airbus Industrie.[46] It will also allow the company to satisfy the demand of many of its customers that it modernize several models of aircraft. "The manufacturer will (henceforth) be striving to build the models the airlines want instead of those it thinks they want," notes an analyst in this field.[47] Others see in this decision the beginning of troubling changes. In moving closer to the logic of the exposed culture, is Boeing not running the risk of sacrificing its long-term development to its short-term commercial objectives, hence jeopardizing its long-standing position as a leader in technology?[48] If confirmed, this evolution would significantly alter the process of innovation within this large sector of the sheltered culture. Over a long period of time, a lower priority for R&D would necessarily mean fewer technological breakthroughs. Loss of the shelter might ultimately lead to slower technical progress in the sheltered culture.

The arrival in January 1988 of a new man to head Boeing—Frank Shrontz—will probably not relieve these anxieties. Unlike his predecessor, Thornton Wilson, an aeronautical engineer who had been involved in some of Boeing's large military programs, Shrontz was trained at the Harvard Business School. He took over as the head of Boeing at a time when the company's future depended on a vigorous commercial and marketing effort. Sales of civilian airplanes had reached a record of $20 billion in 1987, while revenues and profits on the military side were falling as a result of reductions in Pentagon budgets and of a new government policy for military procurement aimed at inducing cost consciousness and greater efficiency among its contractors. Under the leadership of a team better prepared for a marketing approach, Boeing should be able to hold its position in an even fiercer battle against Airbus Industrie.

Two years ago, it was no secret that the US government was deeply concerned about not letting the aircraft industry go the way of automobiles and semiconductors, two sectors that symbolize the invasion of American soil by foreign firms.[49] Hence Washington's pressure on European governments, accused of unfair subsidies to Airbus. In this context, the battle between Airbus and Boeing— two giants of the sheltered culture, each with solid government support—seemed to be in danger of reaching extreme proportions. Fortunately, the unprecedented expansion of the world aircraft market in 1987 and 1988 eased these tensions. Every aircraft manufacturer has been kept busy expanding its production capacity in order to meet growing demand. This does not mean, however, that Boeing has accepted Airbus' penetration of the US market. On the contrary, an even fiercer and broader battle is foreseeable, in a few years, between American aircraft firms and their foreign challengers. The large cuts in defense budgets that seem likely will have a direct impact on the activity of military aircraft manufacturers, and it will be a matter of survival for many of them to convert part of their manufacturing capacity to build civilian airliners.

In other sectors, increased competition has also brought some firms to rationalize their activities. Many firms that were already strongly established in sheltered markets have chosen more shelter, in order to take advantage of abundant military contracts, or to focus their resources on the development of technologies for which they hold a stronger comparative advantage. This is how General Electric came to abandon its consumer electronics division, which it exchanged for the medical imaging branch of the French firm Thomson. GE's strategy provides an example of a company that has decided to concentrate its activities in the sheltered culture. A similar strategic move was made by Honeywell, which in 1986 sold its computer division to Bull and NEC and brought the aerospace branch of Unisys.[50] Honeywell thus abandoned its exposed-culture activities, in which it had never made sufficient profits anyway, and focused most of its attention on the defense sector. Another example is Dupont, which since the mid-1980s has established new ties with the Department of Defense. The purpose of this was to be part of the Pentagon effort in advanced composite materials, as well as to benefit later from government procurement in these materials for the space program or for new aeronautics projects such as the stealth bomber and the hypersonic airplane. According to a company manager in this new sector, Dupont's revenues from advanced materials used in defense and space could reach $1.2 billion by the year 2000.[51]

Projects for the Next Century

As a result of the expanded scope and number of large federal technology programs during the Reagan era, many companies in the exposed culture were attracted to the sheltered culture, or concentrated more of their activities there. Such a diversion of technical and industrial resources should probably be taken into account when we review the causes of the deterioration in America's industrial competitiveness over the past few years. The main cause of this diversion of resources—the government programs themselves—may, however, come under fire in a time of tighter budgetary constraints. Having inherited these programs from the preceding administration, President George Bush will have the difficult task of reestablishing priorities in this domain. Although it is difficult to predict the outcome of the political process that will deal with these choices, several of the elements that will figure in them can be predicted.

One element concerns the tradeoffs between public resources devoted to defense and those spent to reinforce industrial competitiveness. When he took over as Secretary of Defense in October 1987, Frank Carlucci officially confirmed the leveling-off of the defense budget that Congress had already imposed on the Reagan administration the preceding year: defense spending was stabilized at approximately $300 billion a year.[52] For another thing, the debate over federal support of industry was renewed on the occasions of Congress' trade bill and discussion of the Sematech project. The latter cast light on the contradictions in the doctrine responsible for nearly all federal support for technological and industrial development being allocated via the Pentagon. Much pressure was brought to bear on the Secretary of Commerce to have his department rather than the Department of Defense provide federal funding to Sematech. The industries involved were not the last ones to push in this direction. They were worried about seeing the Pentagon once again play the role of grand technological financier and impose upon the project technical and managerial orientations that would not be in the companies' immediate interest.

This subject was far from new, but Sematech provided new ammunition for both the advocates and the opponents of government intervention in commercially oriented technological projects. Over the past 20 years there has been a political consensus in favor of letting the government finance the most audacious projects as long as they contributed to national security and as long as the Department of Defense was ready to handle them. In fact, national-security justification for government involvement in technological development allowed for great flexibility. Along this line, it seemed natural that the Department of Defense should be asked to participate in the funding of Sematech, since almost everybody agreed on its strategic importance. Ironically, though, many Pentagon officials were not at all in tune with this idea. Rather than Sematech, which was aimed at the development of commodity chips, they would have preferred a project focusing on the specific integrated circuits needed by the military. Conversely, the idea that the Department of Commerce might become the official disburser of funds immediatly raised objections even from the less dogmatic among the free-market advocates. Why, indeed, should the government be called upon to shore up what they saw as a business failure, as a lack of foresight by the companies that should themselves have financed a project that was obviously tied to their productive and commercial objectives? Eager to preserve the dogma of government noninterference in industrial affairs in the competitive sector, Secretary of Commerce Malcolm Baldrige emphasized possible problems with GATT when he rejected participation by his department in financing the Sematech project.[53]

This debate is not about to end, since the precedent set by Sematech will undoubtedly be copied. Inspired by the Japanese model, now being applied in Europe as well, cooperative government-industry ventures like Sematech would seem to be a promising approach for strengthening American manufacturing. A sharing of the burden of developing costly new commercial technologies among rival American firms makes sense at this time of intense international competition. Bringing in government resources to speed up this development also makes sense, as long as America's chief competitors are not hesitating to do so. A major difficulty may arise, however, if government rather than industry decides which technologies should be developed. This problem may well affect Sematech as long as the Pentagon is providing the only government support for that project.

Making desirable changes in the process by which the federal government allocates funds for technological development will probably take time, for two main reasons. First, the dogma of government nonintervention in commercial sectors persists. What in Europe is called industrial policy is still regarded as nonsense by many government experts and politicians in the United States. Second, the Department of Defense is still the only government agency with substantial resources at its disposal for technological development.

There is, however, an area where the legitimacy of government intervention is not open to question: fundamental research. The proposal made in 1987 by President Reagan to double the budget of the National Science Foundation by 1992 was probably a political maneuver aimed at quieting the hostility of sectors of the scientific community to large technological programs, be they civilian like the SSC or military like SDI. Even if the more severe constraints that are likely to weigh on the federal budget over the next few years make it less certain that such an objective could be reached, the need for the goverment to significantly increase the funds available for fundamental research has a large consensus in the universities and in industry.[54]

A second important element in the debate over large military technology programs is the evolution of East-West relations. In its quest for budget savings, Congress will obviously make substantial cuts in military budgets as further progress is made in disarmament. The strong growth in military programs in the early 1980s was justified by increased tension between East and West. Political changes occurring today in Eastern Europe may bring the threat of an East-West confrontation to an all-time low, hence precipitating a reverse movement. In early 1988, Frank Press, president of the National Academy of Sciences, said that military budgets should continue to fall during the next few years.[55] In this context, hard choices will be inevitable. Although Ronald Reagan's great projects—SDI, the hypersonic airplane, and other advanced military technologies—were compatible with the budget as long as they were in their low-cost preliminary phases, they can hardly reach full development within a shrinking military budget. More fundamentally, the rationale for bringing most of this new military arsenal to an operational stage will be seriously questioned if the political disintegration of the Warsaw Pact is confirmed.

In this context, what will become of the space program? Aside from the human tragedy, the *Challenger* mishap was perceived as an insult to America's pride, which was all the harder to face in view of the Soviet Union's achievements with its orbiting space station. Three years later, the American sheltered culture has successfully gathered its forces (although at a very high cost) to erase the consequences of the tragedy. But the future of manned flight in the US space program is still under debate, and it is likely that greater emphasis will be given to automated missions.

In spite of the criticism to which the space-station program has been subjected, and in spite of its greatly increased price (the most ambitious version could cost $30 billion), this project may eventually be implemented. It will then keep NASA busy for most of the coming decade. With the space station, the American and the international scientific community will have a very costly but powerful new tool for astronomical and environmental observation of the Earth. Increasing this area of knowledge at a time when it has been discovered that the Earth's ecological equilibrium has been endangered by two centuries of human industrial development should be a major priority for space research during the next few years.

But what about more grandiose schemes, such as returning to the moon or venturing to Mars? On February 17, 1987, the White House for the first time mentioned a manned flight to Mars as a major long-term objective of the American space program.[56] Of course no one today would dare suggest a calendar for such a mission, which would necessarily involve many intermediate steps. In the report she submitted to the NASA Administrator in August 1987 on the goals of the American space program, the astronaut Sally Ride said that she felt it would be possible by the year 2000.[57] But many scientific and technical problems remain to be solved before such a trip can be considered: the development of more efficient propulsion, a vehicle design able to use atmospheric braking when approaching Mars and when returning to Earth, and above all medical progress to overcome the physiological and psychological difficulties of a trip that would last nearly three years. International cooperation would be mandatory, if only for financial reasons.

Congress is obviously not ready to approve a Mars project. A survey made in early 1988 showed that although the public was favorable toward such a space adventure, it was not ready to give it top priority as it was during the glory years of the Apollo program.[58] Consequently, the question of a journey to Mars is premature for the time being. But the Bush administration can be expected to define new and more ambitious long-range goals for space technology, and those decisions are being awaited by scientists, engineers, and entrepreneurs throughout the sheltered culture, who were recently seized by doubt about the value of their objectives and missions. The American sheltered culture needs technological challenges, even if they are less grandiose than they used to be.

A New Challenge for Europe

12

Europe has accumulated several handicaps in the race for technological development. It does not, like Japan, have an industrial
system adapted to the logic of the times. Its dominant logic, even
more than in the United States, is still the logic of the sheltered
culture. In addition to the Japanese challenge, which by now has
become familiar to them, the Europeans are soon likely to be
confronted with a new American challenge. This American challenge will not be, as that of the 1960s was, the result of a technology
gap.[1] Whereas in the 1960s Europe suddenly became aware of its
lag in R&D and high technology, today Europe no longer has
reason to feel inferior to the United States in most areas of scientific
and technological knowledge. The new challenge is not to catch up
with the leaders but to transpose know-how to industry more
effectively. The new challenge is to deal with the arrival on European markets of cheaper American products.

Strongly penalized by the continuous dollar revaluation
between 1981 and 1985, American high-technology companies have
been severely hit by increasingly aggressive foreign competition.
Those that have survived have unquestionably adapted to a much
more competitive environment. Over the past two years they have
attained a more favorable position from which to attempt to reconquer foreign markets, taking advantage of the strong stimulus provided by a substantially lower exchange rate for the dollar. This
"new look" American industry clearly intends to take advantage of

the opening in 1993 of the largest foreign market for high-tech products: Europe.

How can this new challenge be faced by the Europeans at a time when technological development is changing in favor of the exposed culture, the area of Europe's greatest weaknesses?

First of all, it is clearly hopeless for the Europeans to try to row alone against the tide. Long made up of protected national markets, the European high-technology industry cannot stand aside from the irresistible changes that are reducing the scope of the sheltered culture. Europe cannot avoid the problem facing American industry today: adapting to the logic of the exposed culture.

This is particularly true in France, where technological development has for too long been in the hands of the government, which has led, as we saw earlier, to an excessive concentration of resources in the sheltered sectors. Some progress is nonetheless noticeable.

While it is still to early to make a final accounting, the cultural revolution started by Bull is exemplary in many respects. Last heir to a long line of unprofitable or marginally profitable French computer companies, Bull was nationalized together with a number of other French industrial concerns when the Socialists came to power in 1982. But to the amazement of many observers who had criticized this new step taken by the government in controlling industry, for the first time in the history of Bull—France's "national champion" in the computer industry—the company's management and not the government was deciding what the company's strategy should be. This decisive change occurred in 1982 in conjunction with another, no less important change: Bull would no longer be the exclusive beneficiary of government procurement. Surprisingly for all those who feared that the company would not be able to adjust to this change, four years after leaving its shelter Bull was able to balance its books. This result is encouraging, even if the company's long-term viability remains to be seen. Its success would seem to belie my earlier discussion of the difficulties entailed in switching cultures. Bull has the merit of being a pioneer in such cultural change. Let us note, however, that the government's withdrawal from Bull's decision-making process does not mean an absence of government intervention. It only means the abandonment of long-standing sheltered-culture logic.

The presence of government aid can be compatible with the logic of the exposed culture. Aid and shelter should not be confused. Aid is temporary; shelter is permanent. Shelter is public procurement, or its equivalent: government involvement in creating various forms of permanent protection for one or more national champions. Aid—which most often takes the form of direct or indirect government participation in R&D—does not necessarily imply government intrusion into a company's management. This can be seen in the Japanese example. The Japanese government is a master at organizing efficient coordination among companies that have the benefit of public research funds but whose competitiveness it is important not to stifle. The Sematech project in the United States illustrates another model of cooperation, in which firms in the same sector—with the government's help—are creating a pool of resources for the development of new manufacturing technologies. The Sematech project's investment cost—$1.5 billion over five years—shows how much is at stake. It also shows the path that the European semiconductor industry should follow if it wants to keep up with its foreign competitors. This example should be applicable to many European industries in the exposed culture, which have an interest in sharing the costs of developing the technologies of the future. European governments may also play an active role here, by providing financial support for such projects (as is done in the United States and Japan). The Esprit and Eureka programs have already paved the way toward this approach.

Launched in 1984 by the European Economic Community, the Esprit program was intended to encourage pre-competitive research among European firms in the electronics and computer industries. Each project was required to involve at least two companies from two different countries, and was co-financed by the EEC and the companies involved. Esprit is an unquestionable success, having contributed to the development of 220 projects involving more than 500 European companies and laboratories.[2]

The Eureka program, launched in 1984, has followed a similar pattern, although it differs from Esprit in three major respects. First, Eureka's purpose is not only research but also product development aimed at the marketplace. Second, Eureka is not limited to electronics but covers most high-tech sectors. Third, subsidies for Eureka projects come not from the EEC but directly from European governments, which finance the involvement of their own companies. For example, if a Franco-Italian Eureka project is launched in computer graphics, the French and Italian computer firms will receive from their respective governments up to 50 percent of the funds needed for R&D work on the new product they have jointly decided to develop.

At the beginning of 1988 Eureka had contributed to launching 160 projects, involving 1,000 companies, for a total cost of $2 billion. As important as the results has been the new frame of mind created among traditionally hostile European competitors by this cooperation.[3]

Another area where government initiatives might have an impact on the exposed culture in Europe is in encouraging the creation of new firms, an important matter in the long-term renewal of European industry. Although there has been some progress, the number of new companies created in Europe—and particularly in advanced technology—is significantly lower than in the United States. America has two advantages in this area that would be worth looking to for inspiration. The first is the right to be wrong. Failure is a normal thing for the American entrepreneur. It is not borne as a stigma, as it still is in many European countries and particularly in France. The second is a tax policy that, until the 1987 tax reform, was more favorable toward risk than most of the European policies. Lowered capital-gains taxes played a decisive role in the unprecedented increase in the amount of capital available for high technology in the early 1980s in the United States.

Another important ingredient for the growth of small firms is venture capital. Although the amount of venture capital invested in Europe is only a third of that in the United States, the European amount is growing at the rate of 20 percent per year, which is a sign of a new growth dynamic based on small or new companies in Europe. In France—the third-ranking country in the West, but far behind the United States and Great Britain in the investment of venture capital—the volume of funds invested in venture capital has increased by 50 percent per year since 1986.[4] A major impediment to the development of venture capital in Europe has been the fragmentation of markets. If the risks associated with technology-oriented projects are similar on both sides of the Atlantic, capital gains resulting from success are much lower in Europe because of its smaller markets. This problem might lessen, however, after 1993.

These new developments in the exposed culture should not, however, be cause for Europeans to neglect their long-standing comparative advantage: the technologies of the sheltered culture. The sheltered culture is Europe's strength even more than it is America's. Alongside aeronautics and space, two sectors that will continue to require growing government backup to finance ever more costly technological developments, cooperation between Europeans should also take place in other industrial sectors of the sheltered culture. High-speed ground transportation might be one of them. However, there is a precondition for cooperation in this domain: the long-standing quarrels among engineers from neighboring countries must be resolved. While the French used conventional rail technology when designing their TGV, the Germans decided to use magnetic suspension. Close cooperation among large European firms—particularly French and German ones—will be necessary if a large European network of high-speed trains is to be created.

Another promising area for European cooperation is high-definition television (HDTV). The most remarkable thing about this technological development is that it will combine skills from the two technological cultures. On the one hand, implementing HDTV will require satellites, cameras, transmission systems, and a large array of complex technologies that are typical of the sheltered culture. On the other hand, what is at stake with the spread of HDTV is the development of consumer electronics products that will feed a multi-billion-dollar market, mainly for receivers and recorders. But prior to the emergence of these new markets a battle will take place among Europe, Japan, and the United States over the setting of standards. The Europeans recently agreed on a common standard, which has the strong advantage over its Japanese rival of being compatible with today's TV sets. Four large European companies—Thomson, Thorn-EMI, Bosch, and Philips—are cooperating within the framework of a Eureka project to produce a complete HDTV system.[5]

European cooperation in HDTV shows that under the pressure of competition from overseas—in this case from Japan—the Europeans are able to agree on a subject on which they have been divided for a long time: the setting of technical standards for advanced-technology products. Differences in standards strongly contributed in the past to perpetuating barriers between national markets in Europe, for example in telecommunications. In these industrial sectors, the unification of standards is a clear precondition for the unification of European markets that is expected to take place after 1993.

This 1993 goal also means an opening to international competition. In the sheltered culture, deregulation should contribute broadly to this opening. But on this matter, some Europeans are somewhat naive when adopting the credo of the American free-market believers. The American experience shows that, even if deregulation offers substantial advantages, merely deregulating and opening formerly protected markets may not suddenly turn monopolistic firms into highly competitive winners. In fact, just the opposite may occur. A cautious opening of European markets to international competition should be the order of the day, to avoid flooding by foreign products. In some sensitive sectors it would be wise, taking into account the US government's mistakes, to negotiate reciprocity in exchanges before opening previously sheltered European markets to foreign products, particularly those from Asian countries.

Unfortunately, European governments are still far from a unified view on trade policy, particularly concerning imports of industrial products. The best example is provided by the position of the European countries today with respect to automobile imports from Japan. While in some countries these imports are strongly limited by quotas (3 percent of the domestic market in France, 10 percent in Great Britain), they are not subject to any limitation in Germany, Greece, or Ireland. The result is a wide difference in the market share for Japanese cars, ranging from 3 percent in France to 43 percent in Ireland. The subject is even further complicated by the existence of joint ventures between some Japanese and European car makers. For example, should the Bluebird, a car made by Nissan in England, be considered British or Japanese when entering the French market? A bitter dispute is still underway on this matter between the British and French governments.[6]

This example gives a feeling for how difficult the negotiations among Europeans will be when they seek to adopt common rules to control imports of high-tech products. But far tougher negoti-ations will have to be held with the United States and Japan. In Washington and Tokyo, the great 1993 opening of the European market is seen alternately as an opportunity and a threat. Will Europe in 1993 effectively be the world's largest consumer market, wide open to international competition, as many dream, or will it be an impregnable fortress against non-European firms? The reality will probably fall somewhere in between. In spite of the foreseeable difficulties that European governments will have in agreeing on common trade policies and strategies for technological develop-ment, they are hopeful that the great 1993 market will be, more than anything else, a means to strengthen European industry.[7] It is also in the US government's interest to have a stronger European partner. Greater vitality in European industry will, in turn, mean new opportunities and joint ventures with American and Japanese companies. But the Europeans' openness toward such new ventures will depend heavily on the attitudes of Europe's major trading partners—the American and Japanese governments. On this subject, what is happening today in Washington and Tokyo is a matter of more concern than satisfaction. In the United States there is a clear trend toward more protectionism, particularly in Congress, while the often-announced opening of Japan's markets to Western prod-ucts is yet to be seen.

Conclusion

13

The Mojave Desert is not the most hospitable place in the world. Amazingly hot in summer, it is bearable only during the winter, when it attracts millionaires to the sun in artificial oases such as Palm Springs. But the Mojave Desert is also the legendary land of the pioneers of American aviation. It was here, near Edwards Air Force Base, that Charles Yeager first broke the sound barrier in October 1947. It was here as well that Scott Crossfield, flying an X-15, reached Mach 7—seven times the speed of sound—a record that has never been beaten. Edwards is also the landing site for space shuttles.

On December 23, 1986, a strange-looking aircraft landed at Edwards. Its wingspan was as large as that of a Boeing 727, but empty it weighed less than a small automobile. Richard Rutan and Jeana Yeager, the pilots, were cramped in the small cockpit. Nevertheless, they had just spent 9 days, 3 minutes, and 44 seconds in it. *Voyager* had just ended a trip that would go down in aviation history: it had flown around the world nonstop without refueling.

This astonishing project had been born five years earlier during a conversation between Richard Rutan and his brother Burt at a table in a Mojave restaurant. Burt is a talented aeronautical engineer with a passion for designing light aircraft. Richard, a former fighter pilot, was then working in his brother's small aeronautics firm. They decided to develop *Voyager* for their pleasure and glory.[1]

The aircraft was built in a former Air Force hangar near Edwards. Its very construction was remarkable. The most advanced technologies available were used. The airframe was made of carbon fiber and epoxy resin. The two engines, placed one behind the other, served a dual purpose: the first provided power during take-off; the second, derived from a Teledyne Continental prototype developed for the Air Force, provided cruising power with minimal fuel consumption.[2]

On December 14, 1986, Richard Rutan and Jeana Yeager took off from Edwards. Nine days later, at the end of a 25,000-mile flight, they returned to the applause of several thousand sightseers and enthusiastic supporters. They were exhausted but exhilarated to have become a legends of aviation.

As much a fantasy-fulfillment undertaking as a unique accomplishment, Voyager's journey epitomizes the deep American passion for individual accomplishment mixing technological performance and dreams. *Voyager* is a symbol of one of the two technical cultures of America. The triumphant reception and the media coverage given to the pioneers bear witness to its significance. President Reagan saluted their accomplishment.

But is there still a place for that America in a world where competition has become increasingly pitiless? Is there still room for dreams in the cold and disciplined world of "Japanese-style" high tech, where the logic of the exposed culture—bent on the conquest of world markets—dominates?

There are two ways to look at the reality of the American economy at the end of the 1980s.

The first, popular among the European left, is partly supported by some of the observations made earlier in this book. As the preserve of the sheltered technical culture, American industry was for decades able to dominate the world as long as it did not face any serious challenges. This America—in spite of popular beliefs and some evidence supporting them—cannot be portrayed as a land of unmitigated competition, symbolized by flamboyant entrepreneurs riding high technology in a free-market economy. Here, "free market" has often meant domination by monopolies and oligopolies. Public procurement and sheltered markets have provided indispensable protection for the constant drive toward higher performance that characterized this universe—a universe that is today being overwhelmed by the irresistible logic of the exposed culture. Hence the irreversible decline into which American industry would seem to have fallen since the early 1970s—the very historical period in which worldwide competition began. America's progressive loss of stature as an industrial power (a consequence of this expanded competition), its irreversible shift to a service economy, and its transformation into a financial powerhouse whose influence was subsequently overshadowed by Japan's would seem to repeat a pattern of decline previously experienced by Great Britain. Remember that a declining Great Britain still ranked among the world's leaders in scientific and technological development during the first half of the twentieth century. But Britain, like America of today, was unable to take advantage of this valuable resource. The disappearance of large segments of US industry to the benefit of Asian countries would appear to be early evidence of a similar loss of America's influence as an industrial power.

Recent indicators would seem to confirm this gloomy forecast. The persistence of the trade deficit in spite of favorable monetary readjustments appears to be a symptom of economic sickness, revealing an America incapable of reversing its declining competitiveness, pursuing the dream of the affluent society while not yet aware of the need to adjust its consumption to its actual wealth, and thus continuing to live beyond its means at the expense of the rest of the world.

For this America, technology would forever stand as a mirage. The illusion of the country's now-lost power would still be maintained by individual scientific and technical accomplishments (American researchers are not through collecting Nobel Prizes) and by collective technological performance (the conquest of space has only begun). But the country would little by little abandon the real new frontier of technology, the one that produces the vitality and wealth of modern nations. The historical end of the Great American Era would appear to be the logical consequence of the rise to power of a new zone of economic expansion in the Far East. Japan (which in the late 1980s became the world's foremost creditor) and the newly industrialized countries of the Pacific Basin appear to be the countries best adapted today to the imperatives of production and the world of the exposed culture. They seem to have entered the cycle of prosperity that ultimately leads to economic domination. Can a world ruled by information technologies be anything other than a world dominated by Japanese economic logic and discipline?

In his book *Fin de Mondes*, the French economist Christian Stoffaës finds support for this thesis of the irreversible decline of the United States in an analysis of the long cycles of the world economy as they relate to the emergence of the successive waves of innovation.[3] Using the hypothesis of the Russian economist Kondratieff, who attempted at the beginning of the twentieth century to identify long-period cycles in economic history that lead to alternate periods of crisis and prosperity, Stoffaës sketches a plausible scenario in which American hegemony would fade during the final phase of the Kondratieff fourth cycle—a phase that has already begun. The emergence of a new technical system based on information technologies, which we are now witnessing, is to Stoffaës one of the early signs of the birth of the next cycle, which is likely to be dominated by countries in the Pacific Basin.

There is another way to look at American reality, however. That approach is based on a number of observations I made in the field and through direct contact with American industry. Some of these observations, reported in earlier chapters, suggest that American industry is not declining but is in transition, probably in the process of adapting to a world that is radically new to it.

For American high-technology firms, the world of the 1960s was a world of dominance and ease. Isolated from international competition, most of America's high-technology markets at that time were subject to the logic of the sheltered culture, whether they involved emerging technologies, for whose mastery new companies were competing intensely via technology, or whether they involved more mature technological markets dominated by a few producers. This model of competition was also in place outside the United States, where American high-technology multinationals were able to carve out empires with little effort because of their superior technological skills. The clearest examples are IBM in computers, ITT in telephone systems, and Boeing in aeronautics. Now that Europe and Japan have filled part of the gap in high technology, markets are global and price competition is intense. As the world's largest high-technology market, and the one most open to foreign competition, the United States was bound to become the arena for these new battles. The result is the clearly evident retreat of American industry on many fronts. It would be myopic, however, to stop at this first level of observation. Behind the lost influence of

American firms there are clear signs of adaptation. I reviewed them in chapter 11, but the time has now come to reexamine them in a more global perspective, at state and national levels.

At the state level, the last twenty years show a twofold change in comparison with earlier periods. The first change is due to the fact that the United States seems to have ceased to be a country of great domestic migrations such as those described by John Steinbeck. Even though there is still a great deal of mobility in the United States, and although the population is growing faster in some regions of the more prosperous "Sun Belt" as a consequence of new immigration from Asia and Latin America, the population has stabilized in the old industrial areas of the "Rust Belt." The stabilization has come about because of a desire, new in the United States, to cling to one's roots. This aspiration was in part satisfied through policies of industrial conversion. Congressmen and governors elected in the states hit by the decline of traditional industries have refused to bow to the inevitable, and have fought to create or attract new firms to their regions. The result has been a new feeling of regional identity, and the appearance of new forms of local and regional solidarity.[4]

These changes are related to a second phenomenon that has emerged at a regional level over the past ten years: the appearance of new areas of development of high-technology industry in some twenty states, particularly in the "Rust Belt." The cooperative attitude of the new high-tech entrepreneurs in Ohio and Pennsylvania, who are creating companies in "incubators" provided by the nearby universities, is singularly at odds with the blatant individualism of their Silicon Valley counterparts. Nurtured by local financial partners who are sensitive to the communitary wealth represented by the new firms, this new breed of entrepreneur thinks not only in terms of profit but also in terms of job creation. Such is the more friendly kind of high tech that is now taking root in the old industrial regions on the East Coast and around the Great Lakes. These regions also appear to be in the best position to take advantage of the spreading revolution in manufacturing techniques. In addition to their reservoirs of intellectual resources and entrepreneurship, these areas possess an even scarcer asset: they know how important manufacturing is.

At a national level, the United States holds two trump cards in its competition with Europe and Japan. The first is tied to the privileged position of the dollar. As the cornerstone of the world monetary system, the dollar will continue to afford America the privileges attached to a reserve currency. By allowing the dollar's exchange rate to decline strongly,[5] the US government has, to a certain extent, forced the rest of the world since 1985 to share the cost of reestablishing the US balance of trade. No other industrialized nation has ever, over a two-year period, seen the prices of some of its industrial exports decline by 50 percent on its major competitors' domestic markets.

The first indications of a recovery of US industry were visible at the beginning of 1988. There was a significant increase in American exports of intermediate industrial goods—transformed raw materials, such as chemicals, textiles, steel products, and wood pulp, that are used in producing equipment or consumer goods. These products are relatively undifferentiated, requiring no effort for the American producers to adapt them to international markets, and there is high demand elasticity with respect to price.[6] Beyond these early signs, the comparative advantages of the United States were becoming visible again in many markets for equipment and consumer goods. Among them were such high-tech sectors as aerospace, computers, and pharmaceuticals, in which the United States had in the past enjoyed a clear comparative advantage.

The second trump card that America holds is rooted in its two technical cultures. We should remember that the Japanese breakthrough in US markets occurred at a very particular moment in the history of technological development. It was a period that followed soon after the historic split in American high tech between "high fashion" and "ready-to-wear." Here it is important to recall the high-tech scene of the early 1960s, when IBM, Texas Instruments, General Electric, Boeing, Lockheed, and others were lavishly financed by the Pentagon and NASA to develop the most advanced and efficient technologies of the time. It was the era of shelters and of "high-fashion" high tech. Then came the era of "ready-to-wear" high tech, with the expansion of markets in the exposed culture. But just as ready-to-wear clothing eventually benefits from creations in high fashion, innovation in the exposed sector has been nurtured by research done in the sheltered culture. However irresistible the pressures for expanding "ready-to-wear" and increasing the domination of the exposed culture of high tech, in the long run there can be little creation and renewal in high tech without the contribution of the sheltered culture, in which America still clearly holds the lead.

Japan has yet to show that it belongs to this restricted circle of masters of high fashion. Desire for success in itself does not make a fashion designer. In order to innovate, Japanese companies almost always need a foreign source, such as a patent or a product created by technological high-fashion designers in America or Europe. Then the Japanese talent for step-by-step improvement rather than spectacular breakthroughs comes into play, transforming the original custom-made product into a mass-produced commodity manufactured efficiently and cheaply. But Japan on its own cannot yet fully dominate the world of high tech.

In contrast with Japan, a more creative Europe has often given birth to technological breakthroughs but has neglected day-to-day improvements that make products cheap enough for a large number of consumers. Consequently, European consumer markets have been flooded with products made in Japan and Southeast Asia. Because of its cultural heritage, including a tradition of research and creation passed down through generations, the Old World is still one of the major sources of technical high fashion. Similarly, it is in the sheltered culture of high tech that Europe excels.

For the time being, America retains a unique asset: it is the only country that has fully mastered the technologies of both cultures. Both high fashion and ready-to-wear have flourished in American high tech. With Europe, America shares the privilege, still denied to Japan, of controlling the most sophisticated sources of technical knowledge and creation. It shares with Japan the advantage of being efficient—at least in some important sectors—in producing ready-to-wear high tech. America still has the opportunity to exploit its unique advantage and successfully blend the dual talents of its two technical cultures.

This great advantage is, however, unlikely to last forever. In today's technological race, every country attempts to compensate for its weaknesses. For example, during the past few years the European countries have focused on strengthening their skills in the exposed culture. From the Americans, the Europeans have learned how to develop entrepreneurship and venture capital. From the Japanese, they have learned to pool resources to develop new technologies. Indeed, the Eureka and Esprit programs represent a successful adaptation to the specific context of Europe of an important aspect of the Japanese mode of innovation: cooperation between large corporations sharing with the state the costs of technological development while competing for market share. In some sectors, such as semiconductors, American companies have also come to imitate Japanese cooperative strategies. Meanwhile, the Japanese have launched ambitious programs to reduce their lag in high-fashion high tech in such sheltered sectors as biomedicine, booster rockets, and civilian and military aeronautics.

But what superficially seems to be a mutually beneficial learning process across Europe, Japan, and the United States is in fact highly asymmetrical. If the pooling of R&D resources represents significant progress for European as well as American firms, it is only one of the skills needed in the exposed culture. Learning manufacturing efficiency from Japan should also be a high priority. But this may be an unattainable goal, since there is much more than the combination of technological resources and managerial capabilities behind Japan's success. Japan's superiority in ready-to-wear high tech is deeply rooted in Japanese values and culture.

At first sight, the challenge that Japan has to face, which is to develop its influence and its skills in the high fashion of high tech, seems to be just as difficult. It might not be so, though, if we recall the learning process that the Japanese firms have exploited with great success in the past. In the sectors they have targeted, Japanese firms have always chosen to develop first the less sophisticated, easier-to-manufacture products. In order to do so they have had only to buy or to borrow the knowledge needed for this first step, from wherever it was available, and then progressively move up the technological ladder. Perfectly adapted to ready-to-wear high tech, this approach will be more difficult in the high-fashion business because of the greater complexity of products. But there is no reason to believe that, over time, it cannot be implemented with similar success. Here it is important to recall that what characterizes the sheltered industrial sectors is, more than anything else, the fact that the companies belonging to these sectors are driven by technology, and that they manufacture complex custom-made or semi-custom products for a small number of sophisticated customers. The expanding demand for such products is making learning effects more important. For example, this is what is currently happening in the aeronautics industry. The intrusion of Japanese companies into this sector and the implementation of their approach to manufacturing would cause great difficulties for US and European aircraft manufacturers. This has not yet happened, for Japan was, until recently, almost unable to set foot inside the doors of these sheltered markets. But Japanese firms have been getting closer and closer to this goal, with the blessing and often with the direct help of the US government. Indeed, although it is in the interest of neither the Americans nor the Europeans to accelerate the process of facilitating the Japanese companies' entry into the club of high-fashion high tech, this is nonetheless the predictable outcome of current American policies.

For example, the US government has provided Japan with access to key technologies for developing a satellite-launching capability, and has thereby shortened considerably—perhaps by ten years—Japan's learning process in this area, bringing significantly closer the time when Japanese companies will be able to compete directly with European and American companies in launching satellites. The same attitude seems to be prevailing today in the US government with respect to military aeronautics. Japanese-American cooperation on the FSX fighter seems to be a perfect recipe for giving Japan early access to key avionic technologies that will considerably reduce the time Japan needs to be able to compete with Europe and the United States, not only in military aeronautics, but in civilian markets as well. Remember that aeronautics is still an area where many technology transfers take place between the military and civilian sectors. Meanwhile, distressed by Airbus' inroads into the US market for commercial aircraft, American aircraft manufacturers seem to be as slow as their government to identify where tomorrow's threat is likely to arise. For example, Boeing's partnership with Mitsubishi, Kawasaki, and Fuji to produce the 767 has obviously accelerated the learning process of the Japanese aircraft firms.

Indeed, we are approaching the point where Japanese companies will be able to compete with the Europeans and the Americans in aeronautics and space. In doing so, they will introduce into these sheltered industries what they have already been able to master in other sectors of high tech: an efficiency in manufacturing that allows them to progressively transform the most complex products into quasi-commodities.

It might be possible to postpone this new Japanese conquest if the Americans would consider looking to Europe instead of Japan for technological and industrial cooperation. The United States and Europe have a clear common interest, for they are so far the only players in high-fashion high tech. Pooling resources to jointly develop the next generation of technologies could be a way for European and American firms to perpetuate their leadership of the high-fashion club. Morover, models for such cooperation already exist. For example, the CFM-56 jet engine, jointly developed and marketed by SNECMA and General Electric, illustrates how beneficial Euro-American cooperation can be for both parties.

Europe and the United States also have a common interest in pooling resources to counterbalance Japan's superiority in some key areas of ready-to-wear high tech. A good example could be HDTV. Compatible technical standards on both sides of the Atlantic would set the stage for the opening of a large European-American market for HDTV components and products, which should primarily benefit the European and American firms that have been associated with this standard-setting process. The technical and industrial capabilities of US subsidiaries of European companies, such as Philips and RCA-Thomson, could also be an important asset in the development of industrial HDTV products for the American market. Semiconductors seems to be another potential area for cooperation between Europeans and Americans. A sharing of technology and costs between the European consortium Jessi and the American Sematech might increase the likelihood that Europe and America will be able to head off what now seems to be the inevitable domination of Japan in this field.

Such strategic alliances between companies or groups of companies will become more and more common in the future as competition stiffens in most high-tech industries. Obviously, in many cases, instead of a European partner, American companies might prefer to cooperate with a Japanese firm. This might also be the decision of a European company eager to take advantage of a converging interest with a Japanese partner. In the fast-moving world of high-tech markets, where speed in introducing new products is a key criterion for success, there will be a growing need to swap know-how and market shares among companies that will still remain competitors in other markets. Indeed, cooperation and competition are the new conflicting realities of today's high tech. But in this game it is important for each player—and particularly for governments—not to be short-sighted and misled by a short-term advantage. Moreover, it is essential to verify that the same rules are applied by all players, and that none of them is playing with marked cards.

Over the last twenty years the world's technological landscape has changed dramatically. Intense global competition has replaced a world thoroughly dominated by American technology. As a consequence, the United States has lost a substantial share of its influence. But at a time when a rebound seems to have replaced this relative decline of American high tech, while simultaneously the Europeans are cashing in the early dividends of their cooperation in advanced technology, it would be foolish for both the Europeans and the Americans not to draw lessons from the continuous and apparently still irresistible advances of Japan.

In less than two decades, the field of high technology has become more competitive and pluralistic, to the benefit of everyone. It is obviously in everybody's interest that this situation continue to prevail, and governments should actively cooperate toward achieving this goal. In doing so, they should clearly favor policies stimulating international trade and exchanges, but they should also, above all else, avoid circumstances that might allow one of the players to fully dominate the game.

Notes

Introduction

1. "High tech" is a broad term that refers to those industrial sectors using the highest degree of investment in research and development. The ratio between the amount of this investment and the revenues generated varies considerably depending on the branch, from 25.6 percent in aerospace to 0.1 percent in textiles and shoes. Between these extremes are intermediate sectors such as pharmaceuticals (6.4 percent), electrical equipment (4.5 percent), automobiles (2.7 percent), and chemicals (3.0 percent). According to the OECD classification, which I will use as a basic reference throughout this book in speaking of high-tech industries, the first two of these would qualify as high-tech; the latter two would not. Although the cutoff point is necessarily arbitrary, this classification has the merit of isolating six major industrial sectors—aerospace, data processing, electronics and telecommunications, pharmaceuticals, instruments, and electrical equipment—that have several characteristics in common besides their high level of R&D investments.

On the average, these are fast-growing sectors, competing actively on an international scale. High-tech industries provide 31 percent of America's industrial exports and 27 percent of Japan's. They are also characterized by a high rate of innovation and by rapid obsolescence of processes and products. These sectors are generally considered to be of strategic importance by governments. (OECD, *Indicateurs de la Science et de la Technologie*, 1986, pp. 65–69.)

Chapter 1

1. Joseph J. Trento, *Prescription for Disaster* (Crown, 1987), p. 188.

2. Paul B. Stares, *The Militarization of Space* (Cornell University Press, 1985), p. 159.

3. David Dickson, *The New Politics of Science* (Pantheon, 1984), p. 30.

4. Trento, p. 191.

5. Stares, p. 157.

6. See, for example, David Stockman, *The Triumph of Politics* (Harper and Row, 1986).

7. Garry Wills, *Reagan's America: Innocents at Home* (Doubleday, 1987), p. 365.

8. *Venture-Capital Journal: Yearbook 83* (Venture Economics, Inc.).

9. Ronald Reagan was born in Illinois.

10. On this subject see, for example, Everett M. Rogers and Judith K. Larsen, *Silicon Valley Fever: Growth of High Technology Culture* (Basic Books, 1984). This book provides a good analysis of the state of mind of Silicon Valley entrepreneurs. On this subject see also Gene Bylinksy, "California: Great Breeding Ground for Industry," *Fortune*, June 1974, and John Levine, "Living the Dream," *Venture*, March 1983.

11. Rogers and Larsen, p. 223.

12. Ibid., p. 271.

13. January 25, 1983.

14. Jim Bartimo, "Reagan Lauds High Tech Heartland," *Computer World*, January 13, 1981.

15. Department of Defense, "The FY Department of Defense Program for Research, Development, and Aquisitions, Statement by the Undersecretary of Defense, Research and Engineering, to the 99th Congress, First Session, 1985."

16. Samuel F. Wells and Roberts Litwak, *Strategic Defense and Soviet American Relations* (Ballinger, 1987), p. 5.

17. Paul E. Gallis, Mark M. Lowenthal, and Marcia S. Smith, "The Strategic Defense Initiative and US Alliance Strategy," Congressional Research Service, Library of Congress, February 14, 1985, p. 4.

18. William Broad, *Star Warriors* (Simon and Schuster, 1985), p. 100.

19. The father of the project was Peter Hagelsteen, an O Group researcher.

20. Broad, p. 116.

21. *Aviation Week and Space Technology*, February 15, 1981.

22. *New York Times*, March 3, 1985.

23. The Heritage Foundation is an institute devoted to organizing seminars and financing study and research programs. They may cover broad issues such as strategic matters or international affairs.

24. *New York Times*, March 5, 1985.

25. President Reagan, address on national security, March 23, 1983.

26. Trento, p. 196.

27. Ibid., p. 198.

28. US Congress, Office of Technology Assessment, "Civilian Space Station and the US Future in Space," November 1984.

29. Dick Kirchten, "Science Community Still Skeptical of Reagan Science Adviser Keyworth," *National Journal*, August 25, 1982, p. 1633.

30. Trento, p. 202

31. Interview with George Keyworth, June 1986.

32. Vannevar Bush et al., "Science, the Endless Frontier: A Report to the President on a Program for Postwar Scientific Research" (July, 1945; reprinted 1960), NSF 60-40.

33. Harvey Brooks, "Science Indicators and Science Priorities," in Marcel C. La Folette, ed., *Quality in Science* (MIT Press, 1982), pp. 1–32

34. Harvey Brooks, "National Science Policy and Technological Innovation," in R. Landau and N. Rosenberg, eds., *The Positive Sum Strategy* (National Academy Press, 1986), p. 124.

35. Jean-Claude Derian and Bernard Liautaud, "Récession dans la high tech américaine: Crise conjoncturelle ou adaptation structurelle?," note no. 735, Scientific Mission, Embassy of France in the United States, Washington, DC, December 5, 1985.

36. *Hambrecht and Quist Weekly Report*, May 15, 1987.

37. Michel Robin, "Une évaluation du venture capital aux Etats-Unis," note no. 691, Scientific Mission, Embassy of France in the United States, Washington, DC, May 29, 1985.

Chapter 2

1. John Levine, "3000 Sand Hill Road," *Venture*, December 1983.

2. Rogers and Larsen, p. 56.

3. John Wilson, *The New Ventures* (Addison-Wesley, 1985).

4. William J. Perry, "Cultivating Technological Innovation," in Landau and Rosenberg, *The Positive Sum Strategy*.

5. Dick Hanson, *The New Alchemists: Silicon Valley and the Micro-Electronic Revolution* (Little, Brown, 1982).

6. E. Braun and S. McDonald, *Revolution in Miniature* (Cambridge University Press, 1978).

7. Bylinsky, p. 6.

8. *Electronic Business*, August 15, 1986, p. 87.

9. Tom Forester, *High-Tech Society* (MIT Press, 1987).

10. Robert Noyce, "Micro-electronics," in Tom Forester, ed., *The Microelectronics Revolution* (MIT Press, 1980), p. 29.

11. Richard Levin, "The Semi-Conductor Industry," in R. Nelson, *Government and Technical Progress* (Basic Books, 1982), p. 36.

12. Forester, *High-Tech Society*.

13. Noyce, p. 38.

14. Rogers and Larsen, *Silicon Valley Fever*.

15. Hanson, *The New Alchemists*.

16. Gordon E. More, "Entrepreneurship and Innovation: the Electronic Industry," in Landau and Rosenberg, *The Positive Sum Strategy*.

17. Roy Rothwell and Walter Zegveld, *Industrial Innovation and Public Policy: Preparing for the 1980s and the 1990s* (Greenwood, 1981), p. 205.

18. Ibid.

19. Rogers and Larsen, p. 13.

20. Michael Moritz, *The Little Kingdom: Story of Apple Computer* (William Morrow, 1984).

21. OECD, *Les logiciels: l'émergence d'une industrie* (1985).

22. *Business Week*, January 20, 1986.

23. "Microsoft's Gates Uses Products and Pressure to Gain Power in PCs," *Wall Street Journal*, September 25, 1987.

24. *New York Times*, August 10, 1986.

25. T. J. Peters and R. H. Waterman, *In Search of Excellence: Lessons from America's Best-Run Companies* (Warner Books, 1982), p. 258.

26. K. Flamm, *Targeting the Computer* (Brookings, 1987), p. 9.

27. Rogers and Larsen, p. 58.

28. Moore, "Entrepreneurship and Innovation."

29. Rogers and Larsen, p. 274.

Chapter 3

1. *Business Week*, November 15, 1982; *Fortune*, September 26, 1986.

2. John Tirman, *Militarization of High Technology* (Ballinger, 1984), p. 51.

3. J.-M. de Comarmond, "Le programme d'Initiative de Défense Stratégique (IDS): Situation actuelle et orientations de la recherche-développement industrielle et académique," note no. 749, Scientific Mission, Embassy of France in the United States, Washington, DC, May 1, 1986.

4. Seymour J. Deitchman, *Military Power and the Advance of Technology* (Westview, 1983).

5. *New York Times*, April 20 and September 29, 1987.

6. Bill Sweetman, *Stealth Aircraft: Secrets of Future Air Power* (Motor Books International, 1986).

7. *New York Times*, January 25, 1988.

8. Walter A. McDougall, *The Heavens and the Earth* (Basic Books, 1985), p. 123.

9. Ibid., p. 168.

10. Stares, *Militarization of Space*.

11. MacDougall, p. 338.

12. Office of Management and Budget, "The United States Budget in Brief," Fiscal Year 1987.

13. "Will the Aerospace Plane Work?" *Technology Review*, January 1987.

14. Raymond S. Colladay, "Rekindled Vision of Hypersonic Travel," *Aerospace America*, August 1987.

15. NSF, *Science Indicators* (1985).

16. Source: Batelle, Columbus, Ohio.

17. Joan Dopico Winston, "Defense-Related Independent Research and Development in Industry," Office of Senior Specialist, Congressional Research Service, Washington, DC, October 18, 1985.

18. Jacques S. Gansler, *The Defense Industry* (MIT Press, 1980).

19. J.-C. Derian and B. Liautaud, "Le rôle des états dans le développement des industries de pointe aux Etats-Unis," note no. 761, Scientific Mission, Embassy of France in the United States, Washington, DC, 1986.

20. A. Markusen, "Defense Spending and the Geography of High Tech Industries," working paper 423, Institute of Urban and Regional Development, University of California, Berkeley, 1984.

21. Charles M. Tiebout, "The Regional Impact of Defense Expanditures," in Roger E. Bolton, *Defense and Disarmament: The Economics of Transition* (Prentice-Hall, 1966).

22. *Statistical Handbook* (1986), p. 334.

23. *Electronic News*, June 3, 1985.

24. Ronald J. Fox, *Arming America* (Harvard University Press, 1974).

25. Gansler, p. 100.

26. Ibid.

27. Jay Stowsky, "The Impact of Pentagon Policies on the Commercialization of Advanced Technology," BRIE Working Paper no. 17, April 1986, University of California, Berkeley.

28. *Business Week*, June 29, 1987.

29. Fox, *Arming America*.

30. *Wall Street Journal*, October 8, 1987.

31. Ibid.

32. *Business Week*, March 25, 1985.

33. Ibid.

Chapter 4

1. *Wall Street Journal*, May 5, 1987.

2. *Business Week*, November 15, 1982.

3. Forester, *High-Tech Society*.

4. *Business Week*, April 28, 1986.

5. Source: Amgen.

6. Gansler, *The Defense Industry*.

7. Lester C. Thurow, *The Zero-Sum Solution* (Simon and Schuster, 1985), p. 100.

8. Stowsky, p. 29.

9. *Science indicators* (NSF 1985) and AAAS Report no 12 *Red Fiscal Year 82*, no. 87–7, AAAS, Washington, DC, 1987.

10. M. A. Sirbu, T. J. Allen, E. B. Roberts, R. Treitel, "Technological Change and the Traditional Firm: Implications for Government Policies," in *Technologie et avenir régional*, TRP no. 74, Délégation à l'Aménagement du Territoire et à l'Action Régionale, Paris, 1978.

11. William J. Abernathy and James Utterback, "Patterns of Innovation," *Technology Review* 80, no. 7.

12. Rothwell and Zegveld, p. 49.

13. Mary Kaldor, *Baroque Arsenal* (Hill and Wang, 1981), pp. 11–27.

14. Gansler, p. 100.

15. Levin, p. 49.

16. Ibid., p. 43.

17. Braun, p. 88.

18. Levin, p. 44.

19. Norman Asher and Leland Strom, *The Role of the Department of Defense in the Development of Integrated Circuits* (Institute for Defense Analysis, 1983). See also Stowsky, p. 14.

20. Barbara Goody-Katz and A. Philips, "The Computer Industry," in Richard R. Nelson, *Government and Technical Progress* (Basic Books, 1982), p. 163.

21. Ibid., p. 170.

22. Ibid., p. 198.

23. The development of high-tech industries started earlier in the Boston area than around San Francisco. This development followed a significantly different pattern, based on data processing, electronics, and defense-related industries.

24. Kenneth Olson, *Digital Corporation: The First 25 Years* (Digital Equipment Corp., 1983), pp. 8–10.

25. E. Ginzberg et. al., "*Economic Impact of Large Public Procurements*" (Olympus, 1976).

26. Goody-Katz, p. 19.

27. Ibid., p. 212.

28. Flamm, p. 98.

29. David C. Mowery and Nathan Rosenberg, "The Commercial Aircraft Industry," in Nelson, *Government and Technical Progress*, p. 129.

30. Ibid.

31. Ibid., p. 131.

32. Ibid.

33. US Department of Commerce, NTIS, "Competitive Assessment of US Aircraft Industry," Washington, DC, March 1984, pp. 20–24.

34. Mowery and Rosenberg, "Commercial Aircraft Industry."

35. Harold Mansfield, *Vision: The Original Story of Boeing* (Madison Publishing Associates, 1986).

36. *Business Week*, September 23, 1985.

37. Nathan Rosenberg, A. M. Thompson, and S. E. Belsley, "Technological Change and Productivity Growth in the Air Transport Industry," NASA technical memorandum, 1978.

38. Kaldor, p. 88.

39. Mowery and Rosenberg, "Commercial Aircraft Industry."

40. Ibid.

41. *Business Week* (December 1980).

42. Mowery and Rosenberg, p. 114.

43. S. L. Carrol, "Profits in the Air Frame Industry," *Quartery Journal of Economics*, November 1972.

44. Mowery and Rosenberg, p. 113.

45. Leonard S. Hyman, Richard C. Toole, and Rosemary M. Avellis, "The New Telecommunications Industry: evolution and organization," Public Utilities Reports, Inc., Vol. 1, Washington, DC, 1987.

46. At a time when the network was still in early development, the cost of connecting into it was much higher than the price billed to new subscribers.

47. Hyman et al., "New Telecommunications Industry."

48. *IEEE Spectrum*, November 1985, p. 47.

49. Hyman et al., volume 2, pp. 12–13.

50. Braun, p. 33.

51. Ibid., p. 34.

52. Ibid., p. 34.

53. J. Bernstein, *Three Degrees above Zero* (Basic Books, 1987).

54. Ibid.

55. Ibid., p. 10.

56. M. B. Bode, *Synergy, Technical Integration, and Technological Innovation in the Bell System* (Bell Laboratories, 1971), pp. 115ff.

57. Braun, p. 35.

58. Hyman et al., p. 14.

59. Bernstein, p. 85.

60. Bode, p. 127.

61. If one looks at only the industrial production of AT&T, IBM's revenues are 3 times those of AT&T.

62. Brooke and Tunstall, *Disconnecting Parties; Managing the Bell System Break-Up: An Inside View* (McGraw-Hill, 1985).

63. OECD, "Softwares, an emerging industry," 1985.

64. Ibid., p. 47.

65. Ibid., pp. 12, 75.

66. *Business Week*, November 16, 1987, p. 190.

67. *Wall Street Journal*, September 25, 1987.

68. Founded in 1983 by a Frenchman, Philippe Kahan, Borland is now among the world's leaders in microcomputer software packages.

69. *Fortune*, January 17, 1987.

70. Although it is higher than the industry average, this is not an exceptional figure in the pharmaceutical industry.

71. Figures and procedure for the United States. Source: Food and Drug Administration.

72. OECD, The Pharmaceutical Industry (Paris, 1985), p. 27.

73. *New York Times*, December 2, 1987.

74. *Fortune*, January 19, 1987, p. 63.

75. Ibid.

76. See US Department of Commerce, "A Competitive Assessment of the US Pharmaceutical Industry," December 1984. This report shows that the average time between a firm's request and the delivery of marketing authorization varies from 5 months in Great Britain to 23 months in the United States. The process takes longer in only one country, Sweden (28 months).

77. Ibid.

78. Joseph A. Schumpeter, *Business Cycles* (McGraw-Hill, 1939).

79. Raymond Vernon, "International Investment and International Trade in the Product Cycle," *Quarterly Journal of Economics* 80 (May 1966).

Chapter 5

1. Albert H. Teich and Jill H. Pace, *Science and Technology in the USA* (Longman, 1986), p. 10.

2. Richard Barke, *Science, Technology, and Public Policy* (Washington, DC: CQ Press, 1986), p. 33.

3. This section owes a great deal to the discussions I was able to have with Michael Horowitz, Advisor to the director of the Office of Management and Budget, from 1981 to 1985.

4. *Chemical Engineering News*, January 4, 1988.

5. Stephen Strickland, *Politics, Science and the Dread Disease* (Harvard University Press, 1972), p. 150.

6. *Science Indicators*, NSF, Washington, DC, 1985.

7. "The Fairy Godmother of Medical Research," *Business Week*, July 14, 1986.

8. Strickland, p. 34.

9. *Business Week*, July 14, 1986.

10. Strickland, p. 137.

11. Ibid, p. 138.

12. Barke, p. 37.

13. Irvin C. Bupp and Jean-Claude Derian, *Light Water: How the Nuclear Dream Dissolved* (Basic Books, 1978), p. 55.

14. Barke, p. 37.

15. *New York Times*, September 6, 1987.

16. Teich and Pace, p. 7.

17. David Truman, *The Government Progress* (1951).

18. Robert B. Reich, *The Next American Frontier* (Penguin, 1984), p. 40.

19. Ibid., p. 40.

20. Ibid.

21. Bertrand Bellon, *L'interventionisme libéral* (Economica, 1986), pp. 8–94.

22. Reich, p. 93.

23. On the automobile industry see Allans Nevins and Frank Hill, *Ford* (Scribner, 1934). On steel, see Grant McConnell, *Steel and the Presidency* (Norton, 1963).

24. Reich, p. 91.

25. Ibid., p. 178.

26. Ira C. Magaziner and Robert B. Reich, *Minding America's Business* (Vintage, 1983), pp. 215–233.

27. Leonard S. Hyman, Richard C. Toole, and Rosemary M. Avellis, *The New Telecommunications Industry: Evolution and Organization* (Public Utilities Reports, 1987), p. 119.

28. Ibid.

29. US Department of Commerce, National Technical Information Service, "A Competitive Assessment of the US Civil Aircraft Industry," Washington DC, 1984, p. 13.

30. McDougall, p. 138.

31. Paul B. Stares, *The Militarization of Space* (Cornell University Press, 1985), p. 40.

32. McDougall.

33. Harvey Brooks, "Technology and Values: New Ethical Issues Raised by Technological Progress," *Zygow: Journal of Religion and Science*, January 8, 1973.

34. Bupp and Derian, pp. 100–140.

35. The first energy-independence program was launched by Richard Nixon in 1973, but Jimmy Carter considerably increased the energy budget and created the Department of Energy (DOE).

36. Brooks, *National Science Policy and Technological Innovation*, p. 131.

37. Nicholas Wade, "Carter Plan to Spur Industrial Innovation," *Science* 206 (November 16, 1979). See also Frank Press, "Science and Technology in the White House, 1977–1980," *Science* 211 (January 9, 1981).

38. George A. Keyworth, "Four Years of Reagan Science Policy: Notable Shifts in Priorities," Annual AAAS R&D Policy Colloquium, March 1984.

39. *Science Indicators.*

40. Brooks, p. 151.

41. Richard R. Nelson, *Government and Technical Progress* (Basic Books, 1982), pp. 468–470.

42. Ibid., p. 469.

Chapter 6

1. Rogers and Larson, p. 54.

2. Source: CIS.

3. Rogers and Larson.

4. For a description of the American research system, see J. Bodelle and G. Nicolson, "Les universités américaines: dynamisme et traditions," in *Technique et documentation* (Lavoisier, 1985), p. 34.

5. American Association for the Advancement of Science, Annual Report on Research and Development, Washington, DC, 1986.

6. This is only one classfication among many others. NAS, "An Assessment of Research Doctorate Programs in the United States," Washington, DC. 1986.

7. The figure is for 1985. Source: "International Science and Technology Update," NSF, 87–319, 1987.

8. Until 1969 it was called the Instrumentation Laboratory.

9. Dorothy Nelkin, *The University and Military Research* (Cornell University Press, 1972).

10. Ibid., p. 92.

11. Council of Economic Priorities, *Star Wars: The Economic Fallout* (Ballinger, 1988).

12. Interview, January 15, 1986.

13. Ibid.

14. Russel Sabin, "Down (But Not Out) in the Valley," *High Technology*, January 1987.

15. Hambrecht & Quist Weekly Report, May 10, 1985. On this point, see Derian and Liautaud, "Récession dans la high tech américaine."

16. Quick and Finan & Associates, Inc., "The US Trade Position in High Technology: 1980–86," report prepared for the Joint Committee of the US Congress, October 1986, Washington DC.

17. Ibid.

18. G. Keyworth, "The Case for Arms Control and the Strategic Defense Initiative," *Arms Control Today*, 15, no 3 (April 1985).

19. Samuel F. Wells and Robert S. Litwak, eds., *Strategic Defense and Soviet-American Relations* (Ballinger, 1987), p. 12.

20. McDougall, p. 339.

21. Richard Rhodes, *The Making of the Atomic Bomb* (Simon and Schuster), p. 401.

22. The official view on the SDI program is given, for example, in Department of Defense, "The Strategic Defense Initiative," Washington, DC, 1984. See

also James A. Abrahamson, "The Strategic Defense Initiative," *Defense*, August 1984.

23. DARPA experimented in 1985 with a 2–megawatt hydrogen-fluoride laser in the framework of the MID Infrared Advanced Chemical Laser program.

24. The free-electron laser (FEL) at Stanford University has already produced laser pulses of several hundred kilowatts for a 3-micron wavelength.

25. This type of weapon would be based on a beam of charged particles (ions or electrons), accelerated before being channeled or neutralized (as would be required to avoid the effects on the beam of the earth's magnetic field).

26. Among the more representative statements against SDI, see Union of Concerned Scientists, *Empty Promise: The Growing Case against Star Wars*, ed. John Tirman (Beacon, 1986). Among the studies reporting on opposing viewpoints, see the OTA report "Ballistic Missile Defense Technology," US Congress, Washington, DC, 1985. For a viewpoint closer to that of the government, see James R. Schlesinger, "Rhetoric and Realities in the Star Wars Debate," *International Security*, 10, no 1 (Summer 1985). Within Congress, Senator William Proxmire was the most outspoken critic. His position was inspired by Douglas C. Waller and James T. Bruce, "SDI: Progress and Challenges," staff report to Senator William Proxmire and Senator J. Bennet Johnston, US Congress, Washington, DC, 1987.

27. The principal authors of the report gave their conclusions in C. Patel, N. Kumar, and Nicolaas Bloembergen, "Strategic Defense and Directed-Energy Weapons," *Scientific American* 257, no. 3 (September 1987).

28. Ibid.

29. David Parnas, "Software Aspects of Strategic Defense Systems," *American Scientist* 73 (1985), pp. 432–440.

30. Herbert Lin, "The Development of Software for Ballistic-Missile Defense," *Scientific American* 253, no. 6 (1985).

31. On this subject, see also "Strategic Defense Initiative Launch Costs," staff report for Senator William Proxmire, Senator J. Bennet Johnston, and Senator Dale Bumpers, based on the CRS launch cost model, US Congress, Washington, DC, 1987.

32. Schlesinger, "Rhetoric and Realities."

33. John P. Holdren, "The SDI, the Soviets, and the Prospects for Arms Controls," Energy and Resources Group, Berkeley, September 1986.

34. *New York Times*, August 12, 1986. See also "Star Wars' Hollow Promise," *Time*, December 1987.

35. Trento, p. 250.

36. Ibid., p. 288.

37. *New York Times*, June 16, 1981.

38. The physicist Richard Feynman, a member of the Rogers Commission, made a public experiment to prove the effect of low temperature on the O-

ring joints. He gives his opinion in "Mr. Feynman Goes to Washington," *Engineering and Science*, fall 1987.

39. Source: NASA.

40. Clark Evert, "A Dose of Reality for Reagan's Science Spree," *Business Week*, December 15, 1987.

41. The announced cost of the SSC is only $5 billion, but the project is in a very preliminary stage.

42. Evert, "Dose of Reality."

43. *New York Times*, August 12, 1986.

Chapter 7

1. Borrus, Millstein, and Zysman, "U.S.-Japanese Competition in the Semiconductor Industry," Institute of International Studies, University of California, Berkeley, 1982, p. 216.

2. Ibid. See also: "L'acquisition des technologies étrangères par le Japon," CPE Ministère de l'Industrie, des PTT et du Tourisme, Paris, April 1986.

3. Forester, *High-Tech Society*.

4. *New York Times*, June 16, 1985.

5. Borrus, Millstein, and Zysman, p. 50.

6. "High Noon for Fujitsu: Japan's Top Computer Maker Tries to Fight Off IBM and Keep Growing," *Electronics*, May 26, 1986.

7. *Electronics*, December 12, 1986.

8. "La cinquième génération à mi-parcours," *OI Informatique,* January 19, 1987.

9. Gene Bylinsky, "The High Tech Race: Who Is Ahead?" *Fortune*, October 13, 1986.

10. Keichi Oshima, "A High Technology Gap: A View from Japan," in *A High Technology Gap* (Council of Foreign Relations, 1987), p. 100.

11. National Academy of Engineering, *Advanced Processing of Electronic Materials in the US and Japan* (National Academy Press, 1986).

12. Bylinsky.

13. Michel Bernon, "L'industrie américaine des biotechnologies: 1986/87, les années capitales," note no. 745, Scientific Mission, Embassy of France in the United States, Washington, DC, March 25, 1986.

14. J.-C. Derian and B. Liautaud, "La haute technologie américaine: Competitivité ou déclin?", note no. 814, Scientific Mission, Embassy of France in the United States, Washington, DC, February 1987, p. 18.

15. *New York Times*, June 10, 1986.

16. See, in particular, David B. H. Denoon, "Japan and the US: The Security Agenda," *Current History,* November 1983, p. 353, as well as "Science, Technology and American Diplomacy," Fifth Annual Report Submitted to the Congress by the President to Section 563b of Title V of Public Law 95–426, US Congress, April 1984.

17. National Academy of Sciences, *Scientific and Technological Cooperation among Industrialized Countries: The Role of the United States* (National Academy Press, 1985).

18. J.-C. Derian, "Les relations Etats-Unis-Japon dans le domaine de la recherche-développement et des industries de haute technologie," Scientific Mission, Embassy of France in the United States, Washington, DC, 1985, p. 3.

19. See in particular "The Danger of Sharing American Technology," *Business Week,* March 14, 1983, pp. 109–111.

20. "US Blocks Access of Foreign Scientists to High Technology," *Wall Street Journal,* January 25, 1985.

21. Interview with Roland Schmitt, president of the National Science Board and vice-president for research of General Electric, May 1985.

22. Daniel Boy, "La perception de la technologie française par les élites américaines," CETIPOF, Maison des Sciences de l'Homme, Paris, 1987.

23. Pierre Muller, "Airbus: Gérer l'incertitude et la complexité," CERAT, Institut d'Etudes Politiques, Grenoble, December 1987, p. 90.

24. *Wall Street Journal,* September 2, 1987.

25. Muller, p. 48.

26. The launching decision was officially announced at the Bourget air show on June 5, 1987. See Muller, p. 80, and *Wall Street Journal,* May 27, 1987.

27. *New York Times,* May 28, 1987.

28. Source: Intelsat.

29. Ibid.

30. Harvey Brooks, "National Science Policy and Technological Innovation," in *Positive Sum Strategy,* p. 121.

31. Bupp and Derian, *Light Water.*

32. Various sources in the computer industry.

33. Nathan Rosenberg, "Civilian Spillovers from Military Research-Development Spending: The American Experience since World War II," Stanford University, September 1986.

34. *Chemical and Engineering News,* March 13, 1987, 19.

35. Interview with Roland Schmitt, vice-president of General Electric, May 1987, and information from SNECMA, 1987.

36. To say the least, since most of the costs of developing the Titan were assumed by the Air Force.

37. Department of Defense, VHSIC Program, Office of the Undersecretary of Defense for Acquisition, December 31, 1986.

38. Willie Schatz and John W. Verily, "DARPA's Big Push in AI," *Datamation*, February 1984.

39. Leslie Bruechner with Michael Borrus, "Assessing the Commercial Impact of the VHSIC Program," BRIE, University of California, Berkeley, December 1984.

40. Ibid., p. 16.

41. *Fortune*, September 26, 1986.

42. Rosenberg, p. 28. Note that some research on high-performance components may have direct civilian applications. For example, this is the case with gallium arsenide components for use in supercomputers. See *New York Times*, October 14, 1987.

43. Ibid., p. 28.

44. DARPA, "Strategic Computing Initiative Progress Report," Department of Defense, Washington, DC, February 1987.

45. Council of Economic Priorities, "Star Wars, the Economic Fallout." See also Rosenberg, pp. 31–34.

46. See in particular "Progress Is Seen in SDI Research," *Washington Post*, April 19, 1985, as well as "Star Wars at the Crossroad," *Time*, June 23, 1986.

47. See in particular David Hanson, "Defense Technology Base Needs Better Support, Panel Finds," *Chemical and Engineering News*, March 23, 1987; Dwight B. Davis, "Assessing the Strategic Computing Initiative," *High Technology* (April 1985); John P. Holdren and F. Bailey Green, "Military Spending, SDI, and Government Support of Research and Development: Effects on the Economy and the Health of American Science," *Journal of the Federation of American Scientists* (September 1986); see also Stowsky.

48. Rosenberg, p. 30.

49. Samuel I. Doctors, *The NASA Technology Transfer Program: An Evaluation of the Dissemination System* (Praeger, 1971).

50. Derian and Liautaud, "Le rôle des états."

51. *Chemical and Engineering News*, March 13, 1987, p. 19.

52. The Impact of Defense Spending on Nondefense Engineering Labor Markets (Washington, DC: National Academy Press, 1986).

53. John A. Young, *Global Competition, A New Reality: Report of the President's Commission on Industrial Competitiveness* (Government Printing Office, March 1985).

Chapter 8

1. Gerhardt Mensch, *Stalemate in Technology* (Ballinger, 1979). See also CPE, Ministère de l'Industrie et de la Recherche, "Rapport sur l'état de la technique," *Science et techniques*, no. 97 (October 1983), pp. 15–22.

2. Gille Bertrand, *Histoire des Techniques*, La Pléiade, no. 21 (1978).

3. Harvey Brooks, "National Science Policy and Technological Innovation," in *Positive Sum Strategy*, p. 120.

4. Christopher Freeman, "Le défi des technologies nouvelles," in Interdépendance et coopération dans le monde de demain (OCDE, 1987), p. 139. See also John S. Mayo, "The Evolution of Information Technologies," in *Information Technologies and Social Transformation* (National Academy Press, 1985), p. 7.

5. See CPE, "La mutation des télécommunications," in *Rapport sur l'état de la technique*, p. 55–58.

6. US Congress, Office of Technology Assessment (OTA), "Effect of Information Technology on Financial Services Systems," Washington, DC, 1984.

7. Forester, *High-Tech Society*.

8. "Effect of Information Technology on Financial System."

9. US Congress, OTA, "International Competition in Services," Washington, DC, July 1987.

10. *New York Times*, October 28, 1987.

11. Forester, p. 227.

12. US Congress, OTA, "Automation of America's Offices, 1985–2000," Washington, DC, December 1985.

13. Forester, p. 227.

14. James Brian Quinn, "The Impacts of Technology in the Services Sector," in *Technology and Global Industry* (National Academy Press, 1987), p. 119.

15. "International Competition in Services."

16. Freeman, p. 154.

17. Stephen Roach, "Macrorealities of Information Economy," in *The Positive Sum Strategy* (National Academy Press, 1986), p. 96.

18. Daniel Bell, *The Coming of the Post-Industrial Society* (Basic Books, 1973).

19. "International Competition in Services," p. 234.

20. Robert Z. Lawrence, *Can America Compete?* (Brookings Institution, 1984).

21. OCDE, Perspectives économiques de l'OCDE, no. 41 (June 1987). See also J.-C. Derian, "La haute technologie américaine: Ressorts et stratégies," *Futuribles* (July-August 1987), p. 64.

22. Brooks, "National Science Policy," p. 143.

23. Quick Finan and Associates.

24. Source: US Department of Commerce, Washington, DC.

25. *Business Week*, March 3, 1986.

26. Quick Finan and Associates. See also Jean Lemperière, "Quand les Etats-Unis achètent à l'étranger leur matériel de production," *Le Monde Diplomatique* (September 1986).

27. Michaël J. Piore and Charles F. Sabel, *The Second Industrial Divide* (Basic Books, 1984).

28. *Business Week*, March 3, 1986. See also Barry Bluestone and Bennett Harrison, *The Deindustrialization of America* (Basic Books, 1982).

29. Source: SUN Microsystems.

30. Derian, "La haute technologie américaine: Compétitivité ou déclin?," p. 11.

31. Stephen Cohen and John Zysman, *Manufacturing Matters: The Myth of a Post-Industrial Economy* (Basic Books, 1987).

32. Ibid.

33. Thurow, *Zero-Sum Solution*.

34. *Business Week*, March 3, 1987.

35. *New York Times*, October 26, 1987.

36. *New York Times*, May 7, 1987.

37. *New York Times*, May 17, 1987.

38. Five pharmaceutical products based on biotechnology were introduced on the American market in 1987, including TPA, a Genentech product. Thus 1987 marks the beginning of commercial activity in this sector.

39. Embassy of France in the United States, Service de l'Expansion Economique, *Biotechnologies* (special issue, May 1986).

40. Michel Bernon, "L'industrie américaine des biotechnologies: 1986–87, les années capitales," Washington, DC, March 1986, p. 8.

41. *New York Times*, May 17, 1987.

42. Ibid.

43. *New York Times*, October 25, 1987.

44. *New York Times*, October 20, 1987.

45. According to the Securities and Exchange Commission, which investigated circumstances surrounding the crash, 39.9 million transactions of this type were made automatically on October 19, while arbitrage techniques produced 37.6 million transactions. Computerized transactions represented 14.75 percent of the total for the New York Stock Exchange and 21.1 percent of Standard & Poor's 500 stocks.

46. *New York Times*, October 26, 1987.

47. Ibid.

48. No fewer than five evaluations were made of the circumstances of the crash, in particular by the Presidental Task Force on Market Mechanisms, the US Securities and Exchange Commission, and the General Accounting Office. All of them insist that automatic transactions contributed to accelerating the fall of the stock market.

Chapter 9

1. This indicator illustrates globally relative competitiveness between countries. It includes macroeconomic factors tied, for example, to the exchange rate or domestic demand, as well as comparative advantage factors.

2. For semiconductors and computers, see Borrus, Millstein, and Zysman. For telecommunications, see also Thomas McCraw, *America versus Japan* (Harvard Business School Press, March 1986), p. 69–72.

3. See US Congress, Hearings before the Commitee on Science and Technology US House of Representatives, 98th Cong., June 29–30, 1983, on Japanese technological advances and possible US responses using research joint ventures (Government Printing Office, 1984).

4. Borrus, Millstein, and Zysman, p. 53.

5. Ibid, p. 71.

6. Sautter, p. 83.

7. Borrus, Millstein, and Zysman.

8. McCraw, pp. 134–138.

9. Ibid.

10. Sautter, p. 131.

11. Daniel Okimoto, "Regime Characteristics of Japanese Industrial Policy," in Hugh Patrick, ed., *Japan's High Technology Industry: Lessons and Limitations of Industrial Policy* (University of Washington Press, 1986), p. 92.

12. Daniel Okimoto, in *The Positive Sum Strategy* (National Academy Press, 1986), p. 82.

13. Commissariat Général du Plan, "L'électronique: un défi planétaire, un enjeu: L'Europe," Commission des Echanges Internationaux, December 1986, p. 170. Note also that Japanese telecommunications exports are made essentially in the exposed technologies in this sector, i.e., in consumer goods.

14. Note, however, the size of the NTT research budget ($600 million a year). Nevertheless, the NTT is primarily oriented toward product development and not toward basic research as are the Bell Labs.

15. OCDE, *Indicateurs de la science et de la technologie* (Paris, 1986), p. 20.

16. Ibid., p. 62.

---

17. Miquel Pierre, *La seconde guerre mondiale* (Paris, 1986), p. 20.

18. Okimoto, in *The Positive Sum Strategy*, p. 555, also underlines the difficulty for Japanese firms of producing highly complex technical products. David C. Mowery and Nathan Rosenberg, "The Japanese Commercial Aircraft since 1945: Government Policy, Technical Development, and Industrial Structure," Stanford University, April 1985.

19. This situation is changing rapidly, however, because of the importance of the technological transfers being made from the United States to Japan in aeronautics, space, and armaments. In 1983 *Business Week* expressed concern about the magnitude of these transfers (*Business Week*, March 14, 1983, p. 109–112). Also, note that Boeing subcontracting in Japan has become very important.

20. See, for example, James Fallows, "Japan Playing by Different Rules," *Atlantic Monthly* (September 1987). See also Byron K. Marshall, *Capitalism and Nationalism in Prewar Japan: The Ideology of the Business Elite, 1868–1941* (Stanford University Press, 1967), and Chalmers Johnson, *MITI and the Japanese Miracle: The Growth of Industrial Policy, 1925–1975* (Stanford University Press, 1982).

21. On the subject of Japanese strategies, see in particular Okimoto, *Regime Characteristics*; see also Orit Frenkel, "A Case Study of Japanese Industrial Policy," *Journal of Policy Analysis and Management*, 3, no. 3 (Spring 1984), 406–420.

22. Johnson, *MITI*.

23. Fallows, "Japan Playing by Different Rules."

24. If military research is included, the nuclear energy sector at this time was mobilizing nearly half of public R&D resources.

25. Bupp and Derian, pp. 40–50.

26. OCDE, "Electricité, énergie nucléaire et cycle du combustible dans les pays de l'OCDE, 1987," p. 31.

27. Bupp and Derian.

28. OCDE, "Electricité, énergie nucléaire et cycle du combustible dans les pays de l'OCDE."

29. Source: Aérospatiale.

30. Ibid.

31. Emmanuel Chabeau, *L'industrie aéronautique en France, 1900–1950*, and Pierre Muller (cited above) show that engineering logic that produces "beautiful planes" has in fact dominated the entire history of French aeronautics. The reasoning that led to Concorde was no exception but rather confirmed the rule. Only with the Airbus program did commercial logic become dominant.

32. Source: German embassy, Washington, DC.

33. Françoise Vaysse, "CGE-Siemens: La paix armée," *Le Monde*, September 19, 1987.

34. "Siemens Changing from a Tortoise into a Hare," *Electronics*, June 2, 1986, p. 42.

35. *Le Monde*, September 19, 1987.

36. On the subject of recent changes in the organization of the European telecommunications industry, see OCDE, "Tendances des changements de politiques des télécommunications," Paris, 1987.

37. Commissariat général du Plan, "Électronique," op. cit.

38. Deregulation put an end to the "Yalta Agreement" between the two American telecommunications giants that had left the domestic market to AT & T and the international market to ITT.

39. *Electronics*, June 2, 1986.

40. The US Department of Commerce several times expressed its "concern" to the French embassy at seeing France respond to German pressure to back a solution other than the one, proposed by AT&T, that would be dictated by pure economic logic.

41. Robert Zarader, "L'informatique, une stratégie contrainte par une industrie stratégique," in B. Bellon and J.-M. Chevalier, *L'industrie en France* (Flammarion, 1983), p. 221.

42. J. M. Quatrepoint and Jacques Jublin, *French ordinateurs*, ed. Alain Moreau (1976), pp. 88–95.

43. Ibid., pp. 239–247.

44. Zarader.

45. Carlo de Benedetti, "Europe's New Role in a Global Market," in *A High Technology Gap? Europe, America, and Japan* (Council on Foreign Relations, 1987), p. 72.

46. Note that this is only a rough classification between exposed and sheltered activities, since the telecommunications sector is entirely labeled as sheltered. Thus, it does not take into account the exposed part, including, for example, consumer-premise equipment. It is precisely in this area that the Japanese telecommunications exporters have specialized. A more accurate classification would therefore accentuate only the "exposed" character of Japanese industry.

47. Source: Office of Economic Expansion, French embassy, Washington, DC.

48. Source: Donald Agger.

49. In his book *Les 2000: Comment devient-on un grand patron?* (Le Seuil, 1987), Michel Bauer shows the influence of the "grands corps" on the organization of French industry.

50. Bauer.

51. Bupp and Derian.

52. This refers to Philippe Boulin; the present president of Framatome does not belong to the Corps des Mines.

53. This is personal opinion stemming from two years of experience on the Prime Minister's staff.

54. In the case of nuclear energy it is ironic to note that one of the decisive factors in 1969 for choosing American light-water reactors was their ability to be exported. Twenty years later, Framatome, the French builder, has managed to export only four nuclear plants.

55. This was the case in particular for sales of reactors to Iran and South Africa, and for the Indian and Chinese telephone-equipment markets.

56. This discussion is the result of conversations with several officers in the Direction des Relations Economiques Extérieures in the Ministry of Economy and Finances.

57. This discussion benefits from the invaluable advice of Yves Aureille, commercial counselor in charge of the aeronautics sector at the French embassy in Washington.

58. Muller, p. 22.

59. Ibid., p. 25.

60. Interview with Donald Agger, December 1987.

61. Yves Morvan, "L'aéronautique," in *L'industrie en France*, p. 251.

62. US Department of Commerce, "Competitive Assessment of the US Civil Aircraft Industry," p. 46.

63. *Wall Street Journal*, August 10, 1986.

64. OCDE, *The Space Industry* (Paris, 1985), p. 77. See also Patrick Cohendet and André Lebeau, "Choix stratégiques et grands programmes civils," *Economica*, 1987.

65. OCDE, *Indicateurs science et technologie*, p. 130.

66. Henry Ergas, "Does Technology Policy Matter?," in *Technology and Global Industry* (National Academy Press, 1987), p. 197.

67. Commonalities between these two technological programs—technical sophistication, weakness of economic analyses, and technological development taking place up to advance stage in spite of lack of demand—had been emphasized as early as 1971 in the United States. On this subject, see Eads and Nelson, "Government Support of Advanced Civilian Technology: Power Reactors and the Supersonic Transport," *Public Policy* (Summer 1971).

68. Ergas, p. 196.

69. "Le financement de la recherche, du développement, et de l'innovation en République Fédérale d'Allemagne," note by the science and technology counselor of the French embassy in Bonn, December 1986.

70. OCDE, *Science and Technology Policy for the 1980s*, pp. 37–39.

71. Ergas.

Chapter 10

1. Jean-Marc de Comarmond, "Les centres de super-ordinateurs vectoriels pour la recherche universitaire et la recherche publique civile aux Etats-Unis," no. 793, Scientific Mission, Washington, DC, October 14, 1987.

2. Ibid.

3. In fact it is difficult to consider the NEC promotional venture in Houston a real "sale"; NEC wanted to use it as a showpiece and a precedent to begin selling in the United States.

4. J.-M. de Comarmond, "Les centres de super-ordinateurs."

5. For the conseqences of Chen's departure see "The Next Generation at Cray Research," *New York Times*, February 10, 1988, and "Cray Supercomputer Axed, Superstar Departs," *Science*, 237 (September 25, 1987), p. 1558. Note that much of the know-how still belongs to Cray. This explains Chen's association with IBM, which can provide him not only with the financing he needs but also with its technological backing, indispensable for carrying out the projects he left behind at Cray.

6. Source: Poste d'Expansion Eonomique et Mission Technique pour l'Armement, French embassy, Washington, DC.

7. Hyman.

8. *Chemical Week*, November 18, 1987, 8.

9. Ibid.

10. On this subject see Jacques Tamisier, "La diversification des compagnies téléphoniques régionales," note no. 785, Scientific Mission, French embassy, Washington, July 19, 1987.

11. Hyman, p. 143.

12. Ibid., p. 127.

13. Brooke and Tunstall, p. 13.

14. Hyman, p. 149.

15. On the circumstances of this decision taken in the context of Watergate crisis, see Coll, *The Deal of the Century*.

16. IBM is also very active in this market. In July 1982 it signed a cooperative agreement with MITEL for developing automatic switching equipment; in June 1983 it bought 15 percent of the capital of Rolm for connecting IBM computers to private switching equipment manufactured by Rolm; in February 1984 IBM and British Telecom signed an agreement to build a banking network for electronic transfer of funds; in March 1984 IBM created a joint venture with Merrill Lynch to provide stock market information via satellite to IBM-PC users.

17. Hyman, p. 156.

18. An interesting portrait of Baxter, a key figure in the decision to break up AT&T, is given by Coll (pp. 240ff).

19. Trudy E. Bell, "The Decision to Divest: Incredible or Inevitable?" *IEEE Spectrum* (November 1985), p. 50.

20. Ibid.

21. Hyman, p. 157.

22. Brooke and Tunstall.

23. Hyman, p. 158.

24. Only MCI balanced its books in 1987.

25. *New York Times*, February 15, 1987.

26. *Business Week*, May 19, 1986.

27. *Fortune*, May 25, 1987.

28. *Business Week*, May 19, 1986.

29. *Fortune*, May 25, 1987, p. 43.

30. *New York Times*, February 15, 1987.

31. *Business Week*, May 19, 1986, p. 9.

32. *Fortune*, May 25, 1987, p. 48.

33. Ibid.

34. *New York Times*, February 15, 1987.

35. OCDE, "L'industrie des logiciels," p. 68.

36. *Business Week*, December 22, 1986.

37. Source: *National Telemarketing*, quoted in *Fortune*, May 25, 1987, p. 50.

38. *Fortune*, May 25, 1987, p. 50. Also, in January 1988 AT&T took an equity position in Sun Microsystems, a California computer company that was already using the UNIX operating system.

39. Ibid.

40. Roger Noll and Bruce Owen, *The Political Economy of Deregulation: Interest Groups in the Regulatory Process* (American Enterprise Institute, 1983).

41. "Is Deregulation Working?" *Business Week*, December 22, 1987, pp. 50–55. See also "Les transports aériens aux Etats-Unis," report no. 17, June 1987, from the Poste d'Expansion Economique, French embassy, Washington, DC, as well as the study made by the Department of Commerce, "Competitive Assessment of the US Civil Aircraft Industry," p. 91.

42. Michaël Borrus, François Bar, and Ibraham Warde, "The Impact of Divestitute and Deregulation: Infrastructure Changes, Manufacturing Transition, and Competition in the US Telecommunication Industry," Berkeley working paper, September 1984. A synthetic view of the consequences of telecommun-

ications deregulation is given in "The Geodesic Network 1987 Report on Competition in the Telephone Industry," a study made by the consultant Peter Huber at the request of Judge Green.

43. *New York Times*, October 12, 1987.

44. *New York Times*, February 15, 1987.

Chapter 11

1. Barnaby J. Feder, "Test Facility Offers a Glimpse of the Factory of Tomorrow," *New York Times*, December 17, 1986.

2. On computer integrated manufacturing (CIM), see in particular Jeffrey Zygmont, "Flexing Manufacturing Systems," *High Technology* (October 1986), pp. 22–27; US Department of Commerce, Office of Productivity Technology and Innovation, "Flexible Manufacturing: The New Industrial Revolution," August 1987; "High Tech to the Rescue," *Business Week*, June 16, 1986, pp. 100–106; Jaïkumar Ramachandram, "Post-Industrial Manufacturing," *Harvard Business Review* (December 1986), p. 69.

3. *Business Week*, June 16, 1986, p. 104.

4. Ibid.

5. Derian and Liautaud, "La haute technologie américaine: Compétitivité ou déclin?" pt. 2 "Stratégies d'adaptation," note no. 769, Scientific Mission, French Embassy, Washington, DC, February 17, 1987.

6. Source: Apple.

7. Yves Doz, "International Industries: Fragmentation versus Globalization," in *Technology and Global Industry* (National Academy Press, 1987), p. 96.

8. *Business Week*, June 16, 1986, p. 100.

9. Ibid., p. 101.

10. Jeoffroy Bairstow, "GM's Automation Protocol Helping Machines Communicate," *High Technology* (October 1986), pp. 38–42.

11. "High Tech Drive: GM Man Sets New Trend," *Wall Street Journal*, June 6, 1985.

12. "Reviving the American Factory," *New York Times*, January 2, 1987. See also Robert D. Hershey, "Small Manufacturers Lead Revival," *New York Times*, February 10, 1988.

13. Source: Department of Commerce.

14. Bruce Stocks, "The 21st-Century Factories," *National Journal*, February 13, 1988, pp. 382–385.

15. J.-M. de Comarmond, "Le point sur Sematech: Consortium de recherche sur la production de semi-conducteurs," Scientific Mission, Washington, DC, December 23, 1987.

16. "Bids for Sematech Hike at 11th Hour," *Electronic News*, December 7, 1987, p. 83.

17. "Senate Votes Financial Aid to Semiconductor Consortium," AW & ST, December 21, 1987.

18. Ibid.

19. See in particular "The Defense Science Board on Defense Semiconductor Dependency," Office of Undersecretary of Defense for Acquisition-US Department of Defense, Washington, DC, February 1987. See also J.-M. de Comarmond, "Les risques de dépendence extérieure dans le domaine des semi-conducteurs pour la défense des Etats-Unis," note no. 722, Scientific Mission, Washington, DC, March 17, 1987.

20. Peters and Waterman, *In Search of Excellence*.

21. Thurow, *The Zero-Sum Solution*.

22. Philippe Messine, "Les saturniens: Quand les patrons réinventent la société," Mercure de France, 1987.

23. *Washington Post*, July 24, 1985.

24. *Business Week*, July 11, 1986.

25. *Wall Street Journal*, September 15, 1986.

26. *New York Times*, January 25, 1987.

27. *New York Times*, August 9, 1986.

28. Constance Holden, "New Toyota GM plant is US Model for Japanese Management," *Science* 233, July 18, 1986, pp. 273–277.

29. Ibid.

30. Messine.

31. For a survey of regional industrial policies, see Marianne Clarker, *Revitalizing State Economics*, National Governors Association, Center for Policy Research, Washington, DC, 1985.

32. David Birsh and Suzan McCraken, "The Role Played by High Technology Firms in Job Creations," MIT Program for Neighborhood and Regional Change, 1984.

33. On high tech and and regional development and states' policies in this domain, see Derian and Liautaud, "Le rôle des états"; see also US Department of Commerce, Office of Technology Assessment, "Technology Innovation and Regional Development," Washington, DC, July 1984; Herb Brody, "States Vie for a Slice of the Pie," *High Technology* (January 1985); C. Watkins, "States Programs to Encourage the Commercialization of Innovative Technology," National Governors Association, Center for Policy Research, Washington, DC, December 1985.

34. J.-C. Derian and C. Douineau, "Le rôle des états dans les industries de pointe aux Etats-Unis," pt. 2, no. 788, Scientific Mission, French embassy, Washington, DC, July 30, 1987.

35. Derian and Liautaud, "Le rôle des états," pt. 1, p. 32.

36. Derian and Douineau, pp. 24–31.

37. On this subject, see in particular William C. Norris, "Cooperative Research-Development: A Regional Strategy," *Issues in Science and Technology* (Winter 1985), pp. 92–102. W. C. Norris initiated MCC, the cooperative research center on electronics in Austin, Texas.

38. Suh Nam, "Engineering Research Centers," NSF, 1985.

39. Derian and Liautaud, "Le rôle des états," pt. 1.

40. *High Technology* 7, no. 1 (January 1987), pp. 24–32.

41. National Science Foundation, *International Science and Technology Data Update*, NSF 8-319, 1987.

42. Source: *Quid* (1988) and various US sources.

43. *Wall Street Journal,* September 2, 1987.

44. It was in fact a disguised entrance of Boeing into United's capital in the form of participating loans to meet legal requirements. Since the Air Mail Act of 1934, an airline and a plane manufacturer cannot belong to the same firm. Before 1934 United Airlines was a Boeing affiliate.

45. *Business Week,* September 23, 1985.

46. *Wall Street Journal,* September 2, 1987.

47. *New York Times,* January 27, 1987.

48. Ibid.

49. Comments made by representatives from the Commerce Department to representatives from the French embassy.

50. "Honeywell Beats a Retreat from the Computer Wars," *Business Week,* December 15, 1986, p. 30.

51. Source: Du Pont.

52. Office of Management and the Budget, *US Budget FY 1989,* p. 53.

53. Source: Department of Commerce. GATT is the General Agreement on Tariffs and Trade, regulating international trade, based in Geneva.

54. See Harvey Brooks and Roland Schmitt, "Current Science and Technology Policy: Two Perspectives," Science Policy Seminar Series, School of Public International Affairs, George Washington University, February 1985. See also Hanson, *Chemical Engineering News,* March 23, 1987.

55. Interview with Frank Press, February 25, 1988.

56. Craig Couault, "President Signs Space Policy Backing Lunar, Mars Course," *Aviation Week Space.* A $100 million provision was established in the 1988 NASA budget for the pathfinder project aimed at developing generic technologies for a journey to Mars.

57. Sally Ride, "Leadership and America's Future in Space," report to the National Aeronautics and Space Administration, August 1987.

58. *Time*, February 22, 1988.

Chapter 12

1. Jean-Jacques Servan-Schreiber, *Le Défi Américain* (Denoël, 1967).

2. Lynn Krieger Mytelka, "Les Alliances stratégiques au sein du Programme Européen ESPRIT," *CEPII, Economic Prospective Internationale*, 1er trimestre 1989, p. 5–33. See also "Esprit II: Un pas de plus vers l'Indépendance," *Les Echos*, April 18, 1989.

3. "Eureka News," June 19, 1988, newsletter edited by Eureka Secretariat, Brussels.

4. "Vers un capital risque sans frontières," 1988, Siparex, Lyon. See also Pierre Battini "Capital risque: les règles du jeu" (Editions d'Organisation, 1987).

5. "Super Television: The High Promise—and High Risks—of High-Definition TV," *Business Week*, January 30, 1989. See also Jacques Tamisier, "La Télévision à haute définition: L'occasion pour les Etats Unis de revitaliser leur industrie électronique?", note no. 824, Scientific Mission, French Embassy, Washington, DC, March 1989.

6. "L'automobile européenne face aux ambitions nippones," *Le Monde*, March 21, 1989.

7. "La politique de la CEE à l'égard de ses partenaires," *Le Monde*, November 4, 1988; "La CEE à la recherche d'un nouveau protectionnisme," *Le Monde Informatique*, November 7, 1988, p. 50.

Chapter 13

1. *Newsweek*, December 28, 1986.

2. "Voyager Successes in Historic Flight," *New York Times*, December 23, 1986.

3. Christian Stoffaës, *Fin de mondes* (Odile Jacob, 1987).

4. See, in particular, Bruce Babbitt, "The States and the Reindustrialization of America," *Issues in Science and Technology* (Fall 1984). Babbitt, a former governor of Arizona, was a Democratic hopeful for the 1988 presidential election. He is an outspoken supporter of regional industrial policies.

5. At the end of 1987 the dollar had fallen by 40 percent with respect to the mark and 60 percent with respect to the yen since the peak reached in early March 1985.

6. Source: Poste d'expansion économique, French embassy, Washington, DC, 1988.

Index

Bullet train, 174
Bush, Pres. George, 250
Bush, Vannevar, 23
Bushnell, Nolan, 32
Business ethic, vs. civic ethic, 97–98
Buyer/seller relationship, and competition, 86

Capital-gains tax rate, 27, 258
Carlucci, Frank, 250
Carrol, S. L., 68
Carter, Pres. Jimmy, 14, 103, 105
Carter, Thomas, 212
Cassoni, Vittorio, 221–223
Center for Integrated Systems, 106
Centre Nationale d'Etudes Spatiales, 199
Ceramics, research in, 130
Charles Stark Draper Laboratory, 109–110
Civil Aeronautics Board, 102
Clearing House Interbank Payment System, 152
"Clones," 157–158
Cohen, Stephen, 160
Common Market, 178
Compatibility, policy of, 79
Competition
 foreign, as aggressive, 255–256
 international, 55, 246
 Japanese, as internal, 173–174, 257
 monopolistic, 70
 price, 206
 by technology, 247
Complementarity, between large and small firms, 32, 35
Composite materials, 130
Computers
 and automation, 227–235
 "fifth-generation," 129
 government support of, 60–64
 Japan's invasion of markets for, 232–235
 market for as hybrid, 62–64
 mass distribution of, 209
 needed for SDI, 118–119
 networking of, 80
 and software improvement, 79–80
Computer-assisted design, 227–228

Computer-integrated manufacturing, 228–229
Congress, relations of with president, 90–91
Consensual management, 237
Consumers, and deregulation, 223–224
Cooperative government-industry ventures, 251–254
Cooperative research centers, 106
Coors, Joseph, 19
Corpsards, 191–192
Cost-plus basis, 49
Cray, Seymour, 205–206
Cross-fertilization, 109
Customer-contact networks, 220–221
Customer loyalty, 34

DARPA, 142–145, 147
Data-processing service industry, 78–81, 185–188, 211, 217–221
Davis, Morton, 162
DeBakey, Dr. Michael, 93
Debreu, Gerard, 246
Decentralized computing, 150–153
Decentralized management, 237
Defense budget, 250. See also Department of Defense
Defense contracts, and pork barrel, 95
DeJesus, Tony, 239
De Lauer, Richard, 49
Demand analysis, 56, 220
Demand, structure of, 85, 87
Department of Commerce, 231–232
Department of Defense, 16–17, 46, 50–51, 139–146
 product development for, 54–57
 on technology transfers to foreign countries, 133–134
 and transistor development, 58–60
 and university laboratories, 109–110
Deregulation
 of American telecommunications, 185
 benefits from, 223–226

and opening of European markets, 260
pressure for, 76
Division of Applied Research and Advanced Projects, 142–145, 147
Doriot, Gen. Georges, 61
DRAMs (dynamic random-access memories), 172, 233
Draper Laboratory, 109–110
Drug market, government regulation of, 83
Drugs
generic, 211
over-the-counter, 83

East-West relations, and military programs, 252–253
École Polytechnique, 191
Economic Revitalization Fund, 240
Eisenhower, Pres. Dwight D., 102
Electronic money, 152–153
Electronic systems
development of, 74
increasing importance of, 47–48
technology transfers in, 141–142
Electronics market, European, 183–188
Engineering Research Centers, 242
Engines, aircraft, 64–65
EPROMs (erasable programmable memories), 126–127
Equipment supplies, 76, 225
Esprit program, 257
Eureka program, 257–258
European Economic Community, 201, 257
European Space Agency, 199
Exposed culture, 70
characteristics of markets in, 159
Europe focuses on, 268–269
as foreign to Pentagon technologies, 148
in Germany, 202–203
and government aid, 234–235
and human resources, 147–148
Japan as archetype of, 171–178
and mirage of technology, 123
and over-the-counter drug market, 83

reaction of to European challenge, 138–139
technical progress and expansion of, 208–211

Factories, relocation of, 157–158
Federal Aviation Administration, 67
Federal Communications Commission, 101, 185
Federal Drug Administration, 82–83
Federal government
and American industry, 100–102
budget process of, 89–92
as manager of economy, 99–100
middle-class mistrust of, 96
power of, 96–98
and social justice, 97–98
and technological development, 102–105
Federal Reserve Board, 255
Fermilab, 95
Fin de Mondes (Stoffaës), 265
Fletcher, James, 21, 116
Flexible Integrated Manufacturing Center, 232
Food and Drug Administration, 163–164
Foreign subcontractors, increasing US use of, 158
Foreign students, in US universities, 243–246
Foreign trade, 113–114
"Fundamental analysis" programs, 165

Gallium arsenide, 129, 184, 207
Garwin, Richard, 117
Gary, Elbert H., 100
Gates, Bill, 80–81
General Agreement on Tariffs and Trade, 251
Generic technology, centers for, 242
Geographic areas, industrial characters of, 46–47
Geographic concentration, of high-tech industries, 35–36
Germanium, 59

technological development in, 129–131
Jet engines, development of, 40–41, 65
Jobs, Steve, 32–34
Johnson, Pres. Lyndon, 24
Joint Committee on Atomic Energy, 94

Kaisen, 239
Kaldor, Mary, 66
Kaske, Karlheinz, 184
Kendall, Henry, 117
Kennedy, Pres. John F., 12, 23, 102
Keyworth, George, 22–23
Kilbey, Jack, 59
Kleiner, Eugene, 24–26

Labor-productivity growth rates, 113, 157
Lanzi, Jim, 80
Lasers, 18, 116–117
Lasker, Alfred, 92
Lasker, Mary, 92–93
Lawrence, Robert Z., 157
Learning phenomenon, 30, 172, 269–270
"Less government" wave, 223
Life sciences, US vs. Japanese research in, 129–130
Linvill, John, 106
Lobbies and lobbyists, 91–95
Long-distance telephone service, 71–72
Lowe, William, 33

Machine tools, numerically controlled, 235–236
Mahoney, Florence, 92–93
Mansfield, Sen. Mike, 93
Manufacturing Automation Protocol, 230–231
Manufacturing companies, "hollowing" of, 157–159
Markets
European, 258–259
global, 265
regulation of, 87

Markkula, Mark, 33
Mass production, development of, 96–97
Medical research lobby, 91–94
Megachips, 127
Messine, Philippe, 237
Mettler, Ruben, 52
Microcomputers, 32–35, 154
Microlithography, 233–234
Microprocessors, 28–30, 127
Military space research, 43
Military technology, and civilian market, 139
Ministry of International Trade and Industry (Japan), 105, 171–174
Missile defense, 115–120
MITI, 105, 171–174
Monopolies
government contractors as, 49
and R & D expenditures, 86
Moore, Gordon, 30–31
Mowery, David, 68
Multiple redundancies, 153
Multispectrum photography, 43
Myth of the Post-Industrial Economy (Cohen and Zysman), 160

National Academy of Sciences, 108, 147
National Academy of Engineering, 129–130
National Advisory Committee on Aeronautics, 64
National Aeronautics and Space Administration, 11–13
and Air Force, 21
creation of, 42, 102
Rogers Commission report on, 121
National Atmospheric Research Center, 206
National Bureau of Standards, 104, 227, 231
National Institutes of Health, 89, 91, 93, 104
National Science Foundation, 132, 205, 252
Network corporations, 159

Networks, in telecommunications industry, 85–86
New York Stock Exchange, 153, 165–166
Nixon, Pres. Richard, 12, 17, 24, 103, 105
Norris, William, 205
Noyce, Robert, 28, 30
Nuclear energy programs, 102–103, 179, 192
Nuclear magnetic resonance, 183
Nuclear power lobby, 94–95
Nuclear reactors, 179–180, 201

Office of Management and Budget, 89
Office of Science and Technology Policy, 89–90
Office of Technology Assessment, 22
Olson, James, 218, 221–222
Olson, Kenneth, 61
OPEC, 180
Opel, John, 33
Open system interconnection, 230–231
Optical fibers, 129, 183
Opto-electronics, 129
Outsourcing, 158

Paine, Tom, 12
Parallel data-processing architecture, 144
Parnas, David, 119
Patent protection, and pricing, 83
PBX market, 217
Pepper, Sen. Claude, 92
Perkins, Thomas, 24–26
Peters, Thomas, 34, 236
Pharmaceutical industry, 81–83
Phelan, John D., Jr., 165–166
Piore, Michael, 158
Plan-calcul, 186–187
Pollack, Andrew, 34
Population, stabilization of US, 266
Pork-barrel politics, 94–95
Portfolio insurance, 166
Post-industrial economy, 159–161

Presidential Commission on Competitiveness, 148
President's Commission on the Health Needs of the Nation, 93
Press, Frank, 252–253
Prices
  elastic, 85
  semiconductors, 30
  vs. technical requirements, 84
Product cycles, 86, 159
Product development, risks of, 54–56
Production units, size of, 237
Protection
  and exposed culture, 138
  and sheltered culture, 85–86
Protectionism, US trend toward, 261
Public procurement
  in European computer industry, 187–188
  in Germany, 203

Quality control, 228

Random-access memories, 126
Reagan, Pres. Ronald
  campaign of, 14
  defense priority of, 146, 148
  and renaissance of business ethic, 103–105
  on role of technology, 16, 23–24
  and space program, 12–13, 122–133
  and strategic defense, 19–21, 114–120
Reddy, Raj, 229
Reich, Robert, 97
Research
  and applied industrial orientation, 72–74
  secrecy in, 57
Research and development
  in armaments, 56–57
  European vs. American, 201
  federal concentration of, 44–45
  Japanese, 172, 176
  pooling of efforts in, 269
  self-financed, 141

Wright, Rep. Jim, 232–233
Wyngaarden, James, 89, 91

Yeager, Charles, 262
Yeager, Jeana, 262
Young, John, 12, 148

Zschau, Ed, 15
Zysman, John, 160